Systems & Control: Foundations & Applications

Series Editor
Tamer Başar, University of Illinois at Urbana-Champaign, Urbana, IL, USA

Editorial Board
Karl Johan Åström, Lund University of Technology, Lund, Sweden
Han-Fu Chen, Academia Sinica, Beijing, China
Bill Helton, University of California, San Diego, CA, USA
Alberto Isidori, Sapienza University of Rome, Rome, Italy
Miroslav Krstic, University of California, San Diego, CA, USA
H. Vincent Poor, Princeton University, Princeton, NJ, USA
Mete Soner, ETH Zürich, Zürich, Switzerland;
    Swiss Finance Institute, Zürich, Switzerland
Roberto Tempo, CNR-IEIIT, Politecnico di Torino, Italy

For further volumes:
http://www.springer.com/series/4895

Yury V. Orlov • Luis T. Aguilar

Advanced $\mathcal{H}_\infty$ Control

Towards Nonsmooth Theory and Applications

Yury V. Orlov
Electronics and Telecommunication
CICESE Research Center
Ensenada, B.C., Mexico

Luis T. Aguilar
Centro de Investigación y Desarrollo
de Tecnología Digital
Instituto Politécnico Nacional
Tijuana, B.C., Mexico

ISSN 2324-9749 ISSN 2324-9757 (electronic)
ISBN 978-1-4939-0291-0 ISBN 978-1-4939-0292-7 (eBook)
DOI 10.1007/978-1-4939-0292-7
Springer New York Heidelberg Dordrecht London

Library of Congress Control Number: 2014931236

Mathematics Subject Classification (2010): 93B36, 49G52, 35Q93, 70Q05

© Springer Science+Business Media New York 2014
This work is subject to copyright. All rights are reserved by the Publisher, whether the whole or part of the material is concerned, specifically the rights of translation, reprinting, reuse of illustrations, recitation, broadcasting, reproduction on microfilms or in any other physical way, and transmission or information storage and retrieval, electronic adaptation, computer software, or by similar or dissimilar methodology now known or hereafter developed. Exempted from this legal reservation are brief excerpts in connection with reviews or scholarly analysis or material supplied specifically for the purpose of being entered and executed on a computer system, for exclusive use by the purchaser of the work. Duplication of this publication or parts thereof is permitted only under the provisions of the Copyright Law of the Publisher's location, in its current version, and permission for use must always be obtained from Springer. Permissions for use may be obtained through RightsLink at the Copyright Clearance Center. Violations are liable to prosecution under the respective Copyright Law.
The use of general descriptive names, registered names, trademarks, service marks, etc. in this publication does not imply, even in the absence of a specific statement, that such names are exempt from the relevant protective laws and regulations and therefore free for general use.
While the advice and information in this book are believed to be true and accurate at the date of publication, neither the authors nor the editors nor the publisher can accept any legal responsibility for any errors or omissions that may be made. The publisher makes no warranty, express or implied, with respect to the material contained herein.

Printed on acid-free paper

Springer is part of Springer Science+Business Media (www.birkhauser-science.com)

Preface

The $\mathcal{H}_\infty$-disturbance-attenuation approach to robust control has been an active research area since the 80th of the last century. If confined to linear systems, this approach is fully understood from both the frequency-domain and state-space perspectives [29, 40, 57, 98, 141, 142]. In the state-space formulation, the underlying problem of minimizing the $\mathcal{H}_\infty$ norm of a linear control system is viewed as a differential game of two antagonistic persons, and a solution of the problem is represented in terms of certain solutions of the Riccati equations arising in linear quadratic differential game theory [14].

An appropriate extension to nonlinear systems requires a controller design that would guarantee both the internal asymptotic stability of the closed-loop system and its dissipativity with respect to admissible external disturbances. The standard approach here is to construct a storage function for the closed-loop system that would constitute a Lyapunov function verifying the internal stability [61, 66, 122].

For smooth nonlinear systems, the state-space approach has been developed at nearly the same level of generality as that in the linear case. The existence of a desired controller relates to the existence of suitable solutions of appropriate Hamilton–Jacobi–Isaacs partial differential equations, which replace the aforementioned Riccati equations, and the controller synthesis is associated with these solutions.

Due to the robustness and simplicity of implementation, the nonlinear $\mathcal{H}_\infty$ controllers are widely used in practice. For example, in electromechanical applications [92], they are used when the synthesis is based on the passivity of the closed-loop system resulting from a natural storage function, which consists of the energy. Since complex nonlinear phenomena such as dry friction and backlash are not captured by the passivity-based approach, the practical utility of the approach and the achievable performance remain severely limited.

The nonsmooth $\mathcal{H}_\infty$ approach, proposed in [2] for vector fields of class C^0, admits time-periodic and time-varying settings as well [1, 89], and it is capable of counting for sampled-data measurements [16] and for hard-to-model frictional forces and backlash effects [4, 91]. We develop the latter approach in this book, within the Hamilton–Jacobi–Isaacs framework, whose meaning is augmented with

Clarke's proximal solutions, which are not necessarily differentiable. We invoke planning of motion under the virtual constraint approach [112] to constitute the $\mathcal{H}_\infty$-tracking control of underactuated systems. In addition to the finite-dimensional treatment, the synthesis is extended to the infinite-dimensional setting, involving time-delay and distributed parameter systems.

This book is a research monograph that summarizes the investigation we made in the area of nonsmooth $\mathcal{H}_\infty$ control, starting in the early 2000s. Some related topics are not covered here, as this development is confined to continuous vector fields. No specific study is thus proposed for discontinuous systems, thereby leaving impulsive, variable-structure, switched, and hybrid systems beyond the scope of the monograph. Frictional mechanical manipulators, a servomotor with backlash, an underactuated helicopter prototype, all with incomplete measurements, as well as the state-feedback stabilization of current one-dimensional radial profiles in tokamak plasmas are among other applications illustrated in this monograph. Presenting implementation issues and experimental results makes these applications appropriately complete.

The monograph is intended for advanced graduate students, researchers, and practitioners interested in nonsmooth systems' analysis and robust synthesis. Requiring a basic knowledge in nonlinear systems theory, the monograph is written at an advanced level as an addition to standard textbooks such as Helton and James [61], Isidori [66], and Van der Shaft [123]. An emphasis on nonsmooth systems of class C^0 with special attention to the time-varying/time-periodic case as well as to the infinite-dimensional case distinguishes the monograph from existing books in the area.

This work has benefited from the applications developed in collaboration with Joseph Bentsman (regulation of a coal-fired boiler/turbine unit in the presence of the actuator dead zone), Emilia Fridman (LMI-based stabilization of delay heat and wave processes), Leonid Freidovich, Anton Shiriaev, and Simon Westerberg (virtual constraint approach to orbital stabilization of underactuated systems), as well as with Laurent Autrique, Sylvian Brémond, Oumar Gaye, Emanuel Moulay, and Rémy Nouailletas (current profile synthesis in tokamak plasmas). Their contributions in Chaps. 2, 7, 9–11 are gratefully acknowledged. Also, we wish to thank our colleagues Leonardo Acho and Marlen Meza for technical input on the work.

The work was partially supported by CONACYT grants 165958 and 127575.

Ensenada, Mexico   Yury V. Orlov
Tijuana, Mexico   Luis T. Aguilar

Contents

Part I Introduction

1 Linear $\mathcal{H}_\infty$ Control of Autonomous Systems 5
   1.1   Transfer Functions and Their State-Space Realizations 5
   1.2   Hardy Space and $\mathcal{H}_\infty$ Norm 7
        1.2.1 Computing the $\mathcal{H}_\infty$ Norm of Proper Transfer
                Function Matrices 9
        1.2.2 Computing the $\mathcal{H}_\infty$ Norm of Strictly Proper
                Transfer Function Matrices 10
   1.3   Optimal and Suboptimal Synthesis 10
        1.3.1 Plant Description and Basic Assumptions 11
        1.3.2 Problem Statement 12
        1.3.3 Problem Solution 13
        1.3.4 LMI-Based Solution 15

2 The LMI Approach in an Infinite-Dimensional Setting 23
   2.1   Historical Remarks and Notation 23
   2.2   The Lyapunov–Krasovskii Method 25
   2.3   Exponential Stability in a Hilbert Space 27
   2.4   Exponential Stability of a Delay Heat Equation 31
   2.5   Exponential Stability of a Delay Wave Equation 36

3 Linear $\mathcal{H}_\infty$ Control of Time-Varying Systems 43
   3.1   Preliminaries... 43
        3.1.1 Strict Bounded Real Lemma in Time-Varying
                and Periodic Settings 44
   3.2   Synthesis of Time-Varying Systems.................................. 49
   3.3   Synthesis of Periodic Systems 52

4 Nonlinear $\mathcal{H}_\infty$ Control 55
   4.1   The $\mathcal{L}_2$ Gain and $\mathcal{H}_\infty$ Gain of a Nonlinear System 55

		4.2	Control Objective	58
		4.3	Isaacs' Equation and State-Space Solution	59

Part II Nonsmooth $\mathcal{H}_\infty$ Control

5 Elements of Nonsmooth Analysis 67
  5.1 Semicontinuous Functions and Their Proximal
       Sub- and Superdifferentials 67
       5.1.1 Proximal Sub- and Supergradients
              in a Time-Varying Setting 68
  5.2 Viscosity Solutions of First-Order PDEs 69
  5.3 Nonsmooth $\mathcal{L}_2$-Gain Analysis 71
       5.3.1 Basic Assumptions and Definitions 71
       5.3.2 Instrumental Lemmas 72
       5.3.3 Hamilton–Jacobi Inequality and Its Proximal
              Solutions 74
       5.3.4 Global Analysis 75
       5.3.5 Local Analysis 76
       5.3.6 Periodic and Autonomous Cases 79

6 Synthesis of Nonsmooth Systems 81
  6.1 Synthesis of Time-Varying Systems over
       Continuous-Time Measurements 81
       6.1.1 Basic Assumptions and Problem Statement 82
       6.1.2 Dynamic Nonsmooth Hamilton–Jacobi–Isaacs
              Inequalities 83
       6.1.3 Global State-Feedback Design 84
       6.1.4 Global Output-Feedback Design 86
       6.1.5 Local State-Space Solution 88
  6.2 Local Output-Feedback Synthesis of Periodic
       and Autonomous Systems 94
       6.2.1 $\mathcal{H}_\infty$-Design Procedure in Periodic
              and Time-Invariant Settings 96
  6.3 Local Output-Feedback Synthesis over Sampled-Data
       Measurements 96
       6.3.1 Main Result 97
       6.3.2 Time Substitution-Based Transformation into
              the $\mathcal{H}_\infty$-Control Synthesis via Continuous
              Measurements 98
       6.3.3 Proof of the Main Result 102
       6.3.4 $\mathcal{H}_\infty$-Design Procedure over Sampled-Data
              Measurements 104

7	\multicolumn{3}{l	}{LMI-Based $\mathcal{H}_\infty$-Boundary Control of Nonsmooth Parabolic and Hyperbolic Systems .... 105}
	7.1	Boundary Stabilization of a Semilinear Parabolic System 105	
		7.1.1 Exponential Stability 107	
		7.1.2 $\mathcal{H}_\infty$-Boundary Control 109	
	7.2	Boundary Stabilization of a Semilinear Hyperbolic Equation 112	
		7.2.1 Exponential Stability 114	
		7.2.2 $\mathcal{H}_\infty$-Boundary Control 117	

Part III Benchmark Applications

8 Advanced $\mathcal{H}_\infty$ Synthesis of Fully Actuated Robot
  Manipulators with Frictional Joints..................................... 123
  8.1 Scalar Friction Models 124
      8.1.1 Static Models 124
      8.1.2 Dynamic Models.. 125
  8.2 Problem Statement 126
  8.3 Nonsmooth Synthesis via Dynamic Friction Compensation 128
  8.4 Discontinuous Synthesis via Static Friction Compensation 132
  8.5 Experimental Study 136
      8.5.1 Experimental Setup... 136
      8.5.2 Three-DOF Manipulator Model 138
      8.5.3 Experimental Results 139
  8.6 Synthesis over Nonlinear Sampled-Data
      Measurements: A Case Study.. 144
  8.7 Concluding Remarks 147

9 Nonsmooth $\mathcal{H}_\infty$ Synthesis in the Presence of Backlash 151
  9.1 Output Regulation of a Servomechanism with Backlash 151
      9.1.1 Dead Zone–Based Model of Backlash Phenomenon 152
      9.1.2 Problem Statement and Control Synthesis 153
      9.1.3 Experimental Study... 156
  9.2 Output Regulation of a Coal-Fired Boiler/Turbine Unit
      with Actuator Deadzone 159
      9.2.1 Identified Model of a Coal-Fired
            Boiler/Turbine Unit... 159
      9.2.2 Problem Statement 161
      9.2.3 Nonsmooth $\mathcal{H}_\infty$ Synthesis 164
      9.2.4 Simulation Results.. 165

10 $\mathcal{H}_\infty$ Generation of Periodic Motion of Mechanical
   Systems of One Degree of Underactuation............................. 169
   10.1 Virtual Constraint Approach 170
        10.1.1 Reparametrization of a Motion 171
        10.1.2 Coordinate Transformation............................... 174

		10.1.3	Coordinate-Feedback Transformation	175
		10.1.4	Moving Poincaré Section, Transverse Coordinates, and Transverse Linearization	177
	10.2		Orbital Synthesis Procedure via $\mathcal{H}_\infty$-Position Feedback	178
	10.3		Case Study: A 3-DOF Underactuated Helicopter	179
		10.3.1	Dynamic Model	180
		10.3.2	The Motion Planning	182
		10.3.3	$\mathcal{H}_\infty$ Orbitally Stabilizing Synthesis	183
		10.3.4	Simulation Results	186
11	**LMI-Based $\mathcal{H}_\infty$ Synthesis of the Current Profile in Tokamak Plasmas**			**191**
	11.1		Modeling and Problem Statement	191
	11.2		Proportional State-Feedback Synthesis	195
		11.2.1	Disturbance-Free Stabilization	196
		11.2.2	Disturbance Attenuation	199
	11.3		Proportional-Integral State-Feedback Synthesis	201
		11.3.1	Stabilization Under Time-Invariant Disturbances	201
		11.3.2	$\mathcal{H}_\infty$ Design	204
	11.4		Simulation Results	206

References 211

Index 217

Abbreviations

ARE	Algebraic Riccati equation
ARI	Algebraic Riccati inequality
DOF	Degree of freedom
DRE	Differential Riccati equation
DPS	Distributed parameter system
LMI	Linear matrix inequality
LOI	Linear operator inequality
LQ	Linear quadratic (theory)
LQG	Linear quadratic Gaussian (theory)
LTD(S)	Linear time-delay (system)
LTI(S)	Linear time-invariant (system)
METIS	Minute Embedded Tokamak Integrated Simulator
MIMO	Multi-input–multi-output
ODE	Ordinary differential equation
PDE	Partial differential equation
SISO	Single-input–single-output
Tokamak	Toroidal chamber with magnetic coils (transliteration from the Russian "TOroidal'naya KAmera s MAgnitnymi Katushkami")

Part I
Introduction

Since Zames [139] introduced the basic motivations for $\mathcal{H}_\infty$ optimization, this problem has attracted considerable research interest. Initially, $\mathcal{H}_\infty$-optimal designs, such as [45, 56, 75, 105], to name a few, including that of [43] made in the infinite-dimensional setting, were developed in the frequency domain. The main research tools in the frequency domain were operator and approximation theory, polynomial methods, spectral factorization, and Youla parametrization. We refer the reader to [44] for an extensive literature review and rather comprehensive study in this regard.

In the state-space formulation, the problem of minimizing the $\mathcal{H}_\infty$ norm is viewed as a differential game of two antagonistic persons. The derivation of state-space solutions to $\mathcal{H}_\infty$-optimal control problems, made by Doyle, Glover, Khargonekar, and Francis [38], was truly a breakthrough in linear control theory. Along with other attractive features, the time-domain approach relied on familiar state-space tools from the classical linear quadratic (LQ) and LQ Gaussian (LQG) theory. Compared to the frequency-domain approach, it allowed the reformulation of $\mathcal{H}_\infty$ optimization in the time-varying and finite-time horizon settings [100, 119, 120] as well as in the infinite-dimensional setting [124]. Moreover, significant interest then emerged in extending this derivation to nonlinear systems. The extension was based on the dissipativity property that was introduced by Willems [132, 133] not only for linear systems but also as a general system property, generalizing the notion of passivity in nonlinear electrical networks, and Lyapunov and input–output stability of nonlinear feedback systems. Later on, Isidori and Astolfi [67], Van der Shaft [122], and Ball et al. [13] crucially used the dissipativity concept in the stability analysis and $\mathcal{L}_2$-gain analysis of smooth nonlinear systems, where the importance of the dynamic game theory by Basar and Olsder [15] and Basar and Bernhard [14] was also realized.

It is the time-domain approach that prevails in this book, where the current framework of the nonlinear state-space $\mathcal{H}_\infty$-control theory, summarized in [14, 61, 66, 123], is extended toward nonsmooth systems. Although familiarity with nonlinear systems theory and nonsmooth analysis is assumed, such as that found within the textbooks by Isidori [66] and Clarke [31], the background material, forming Part I, makes the book self-contained. The book consists of three parts and is organized as follows.

Part I outlines fundamentals of the $\mathcal{H}_\infty$-control theory that are relevant to the subsequent nonsmooth development. Chapter 1 is confined to linear time-invariant systems and is the only chapter that touches on the frequency domain. The $\mathcal{H}_\infty$ norm is first defined on transfer matrices, and relations between the frequency- and time-domain approaches to linear time-invariant systems are then established. Standard necessary and sufficient conditions of the $\mathcal{H}_\infty$-suboptimal control problem to have a solution with a disturbance attenuation level $\gamma > 0$ are given in terms of the existence of positive semidefinite solutions of certain algebraic Riccati equations (AREs). With these solutions, the constructive $\mathcal{H}_\infty$ synthesis becomes available, and it appears with a useful separation structure similar to that of the LQG design. A standard approach here is to construct a storage function for the closed-loop system that could also be used as a Lyapunov function for verifying the internal stability.

Due to the well-known strict bounded real lemma, the aforementioned AREs possess positive semidefinite solutions if and only if the corresponding algebraic Riccati inequalities (ARIs) possess positive definite solutions. A particularly nice feature of the latter solutions is that they constitute a strict quadratic Lyapunov function with a negative definite time derivative along the trajectories of the nominal plant as opposed to those of the former equations constituting a nonstrict quadratic Lyapunov function whose time derivative along the same trajectories is only semidefinite. While being quadratic in unknown variables, the ARIs are shown to be equivalent to linear matrix inequalities (LMIs), which allows one to apply, for instance, the MATLAB® LMI toolbox for solving the ARIs. Thus, an alternative $\mathcal{H}_\infty$ synthesis, relying on the positive definite solutions of the ARIs, is also available.

In Chap. 2, the LMI approach, which has long been recognized to constitute a powerful tool for robust control of time-delay systems, is further developed in the infinite-dimensional setting, where a general framework is presented for exponential stability analysis of linear time-delay systems in a Hilbert space with a bounded operator acting on the delayed state. The approach yields an effective stability analysis tool for distributed parameter systems (DPSs) as well; its efficacy is illustrated by scalar linear partial differential equations (PDEs) of parabolic and hyperbolic types. While this chapter falls out of the finite-dimensional background of Part I, it is written for the benefit of readers interested in the $\mathcal{H}_\infty$ synthesis of nonsmooth DPSs, subsequently given in Chaps. 7 and 11. A reader may, however, skip Chap. 2 without any disruption in his or her study of finite-dimensional systems.

Extensions of the $\mathcal{H}_\infty$ synthesis to time-varying and nonlinear systems, governed by ordinary differential equations (ODEs), are presented in the next two chapters. Since the $\mathcal{H}_\infty$ norm, defined on transfer matrices, is not straightforwardly generalized to such a system, its time-domain version is invoked. Once translated into the time domain, the $\mathcal{H}_\infty$ norm becomes the $\mathcal{L}_2$-induced norm of the so-called input–output operator, acting from the input time functions to output time functions for initial state zero. The $\mathcal{L}_2$-induced norm, thus defined, remains applicable to time-varying and nonlinear systems as well, thereby yielding a natural reformulation of the $\mathcal{H}_\infty$-(sub)optimal control problem in terms of this norm.

The direct state-space solutions of the $\mathcal{H}_\infty$-(sub)optimal control problems in the time-varying and periodic settings are successively addressed in Chap. 3. These solutions are derived based on the time-varying and periodic generalizations of the strict bounded real lemma, which are developed first. While being developed in a similar manner as that in the time-invariant case, the resulting time-varying synthesis relies on the differential (rather than algebraic) Riccati equations (DREs) and on their perturbed versions (rather than on differential inequalities). Once applied to a time-periodic system, the proposed synthesis is shown to be periodic as well.

In Chap. 4, the nonlinear $\mathcal{H}_\infty$ synthesis is presented at the same level of generality as that in the linear case. The formulation of the nonlinear $\mathcal{H}_\infty$-control problem is confined to autonomous affine systems (i.e., time-invariant nonlinear systems that are linear in control signals and external disturbances), and it requires a controller design that guarantees both the internal asymptotic stability of the closed-loop system and its dissipativity with respect to admissible external disturbances. The existence of a desired controller is related to the existence of suitable solutions of certain Hamilton–Jacobi–Isaacs PDEs, and the controller design is associated with these solutions. Although the design procedure results in an infinite-dimensional problem, this difficulty is circumvented by solving the problem locally by means of the AREs that appear in solving the $\mathcal{H}_\infty$-control problem for the linearized system.

In Part II, the state-space approach to $\mathcal{H}_\infty$ optimization is further developed in the nonsmooth setting. Chapter 5 briefly presents the Clarke's supergradient calculus to be used in the subsequent $\mathcal{L}_2$-gain analysis of nonsmooth systems. Since the Hamilton–Jacobi PDE or inequality, associated with the nonlinear $\mathcal{L}_2$-gain analysis, may not admit a continuously differentiable solution, the present $\mathcal{L}_2$-gain analysis follows the line of reasoning where the corresponding Hamilton–Jacobi expressions are viewed in the sense of Clarke proximal superdifferentials and are required to be negative definite (i.e., to be in the form of inequalities rather than in the form of equations).

In Chap. 6, the proposed line of reasoning is adopted to derive an $\mathcal{H}_\infty$-design procedure in the nonsmooth setting with no a priori-imposed stabilizability–detectability conditions on the control system. The extra work—formidable in the nonsmooth case—of verifying these conditions is thus obviated. The resulting controller is associated with specific proximal solutions of the Hamilton–Jacobi–Isaacs partial differential inequalities, and it is straightforwardly designed while solving the problem locally. The nonsmooth synthesis is developed in the general time-varying setting and is additionally specified for periodic and autonomous systems with focus on the periodic and, respectively, time-invariant controller designs. Local output-feedback synthesis is presented over sampled-data measurements as well.

The LMI-based $\mathcal{H}_\infty$ synthesis of uncertain distributed parameter systems, governed by nonsmooth PDEs of parabolic and hyperbolic types, is developed in Chap. 7. The nonsmooth uncertainties are admitted to be time-, space-, and state-dependent with a priori known upper and lower bounds. Sufficient exponential stability conditions with a given decay rate are derived in the form of LMIs. These conditions are then utilized to synthesize $\mathcal{H}_\infty$ static output-feedback boundary controllers of the underlying parabolic and hyperbolic systems.

Capabilities of the nonsmooth $\mathcal{H}_\infty$ approach developed are illustrated in Part III by means of several benchmark applications. The presentation here aims to support the theory and emphasizes the control algorithms, whereas presenting the implementation issues and experimental results makes these applications appropriately complete.

Fully actuated frictional mechanical manipulators and servomotors with backlash are studied in Chaps. 8 and 9, respectively, to demonstrate that the proposed synthesis is capable of counting for nonsmooth hard-to-model phenomena such as frictional forces and backlash effects.

In Chap. 10, planning of motion under a virtual constraint approach [112] is invoked to constitute $\mathcal{H}_\infty$-tracking control of a helicopter prototype having three degrees of freedom (3 DOF) and one degree of underactuation.

Finally, the LMI-based $\mathcal{H}_\infty$ synthesis of Chap. 7, developed in the PDE setting, is illustrated in Chap. 11 via the state-feedback stabilization of the current one-dimensional radial profiles in tokamak plasmas.

Chapter 1
Linear $\mathcal{H}_\infty$ Control of Autonomous Systems

The linear $\mathcal{H}_\infty$-control theory makes extensive use of transfer functions, which are revised first. Then the $\mathcal{H}_\infty$ norm is introduced in the Hardy space of stable transfer matrices, and its optimization is given for linear controlled plants in terms of two algebraic Riccati equations (AREs) with a separation structure reminiscent of the classical linear quadratic Gaussian (LQG) theory. Finally, an alternative $\mathcal{H}_\infty$ synthesis, relying on two algebraic Riccati inequalities (ARIs), is developed for later extension to the nonsmooth setting.

1.1 Transfer Functions and Their State-Space Realizations

A generic *rational transfer function* between the single input u and the single output z (SISO) of a linear time-invariant (LTI) system is of the form

$$G(s) = \frac{b_{n_0} s^{n_0} + \cdots + b_1 s + b_0}{s^n + a_{n-1} s^{n-1} + \cdots + a_1 s + a_0}, \qquad (1.1)$$

where n is the order of the denominator (pole) polynomial, also referred to as the order of the system, n_0 is the order of the numerator (zero) polynomial, a_i, $i = 0, \ldots, n$, and b_j, $j = 0, \ldots, n_0$, are the real-valued coefficients of, respectively, the pole and zero polynomials.

Hereinafter, the Laplace transform argument $s = \alpha + j\omega \in \mathbb{C}$ contains the real-valued part $\alpha \in \mathbb{R}$ and imaginary part $j\omega$, with $\omega \in \mathbb{R}$ and $j = \sqrt{-1}$ being the imaginary unit. Since the meaning remains clear, the usual abuse stands throughout for making no distinction between a time-domain signal and its Laplace transform. In this regard, one may realize (1.1) by a state-space realization

$$z^{(n)} + a_{n-1} z^{(n-1)} + \cdots + a_1 z + a_0 = b_{n_0} u^{(n_0)} + \cdots + b_1 u + b_0. \qquad (1.2)$$

For multi-input–multi-output (MIMO) systems with $u = (u_1, \ldots, u_m)^T$ and $z = (z_1, \ldots, z_l)^T$, $G(s)$ is a matrix of transfer functions between particular input and output components. Let an LTI system be given by means of its state-space representation

$$\dot{x} = Ax + Bu,$$
$$z = Cx + Du, \quad (1.3)$$

where $x(t) \in \mathbb{R}^n$ is the state vector, $u(t) \in \mathbb{R}^m$ is the input, $z(t) \in \mathbb{R}^l$ is the output, and A, B, C, D are matrices of appropriate dimensions. Once specified in terms of the state-space data

$$G(s) := \left[\begin{array}{c|c} A & B \\ \hline C & D \end{array} \right], \quad (1.4)$$

the transfer function between the input u and the output z under the trivial initial condition $x(0) = 0$ is expressed in the form of the function matrix

$$G(s) = C(sI - A)^{-1} B + D \quad (1.5)$$

of the Laplace transform argument $s \in \mathbb{C}$, where $I \in \mathbb{R}^{n \times n}$ is the identity matrix. Indeed, applying the Laplace transform to the state-space representation (1.3) subject to $x(0) = 0$ yields

$$sx(s) = Ax(s) + Bu(s), \quad (1.6)$$
$$z(s) = Cx(s) + Du(s), \quad (1.7)$$

and then eliminating the Laplace image

$$x(s) = (sI - A)^{-1} Bu(s) \quad (1.8)$$

from (1.6), (1.7) results in

$$z(s) = [C(sI - A)^{-1} B + D]u(s), \quad (1.9)$$

thereby arguing (1.5).

A transfer function $G(s)$ is said to be *proper* iff its frequency response $G(j\omega)$ possesses a finite limit as $\omega \to \infty$. If, in addition, $\lim_{\omega \to \infty} G(j\omega) = 0$, the transfer function $G(s)$ is then said to be *strictly proper*.

Clearly, a scalar transfer function (1.1) is proper iff $n_0 \leq n$ of the pole polynomial, and it is strictly proper iff $n_0 < n$. In turn, a transfer function matrix (1.5) is proper iff the order of the numerator polynomial is less than or equal to that of the denominator polynomial in every entry of $G(s)$, whereas it is strictly proper iff every entry of $G(s)$ comes with the denominator polynomial of higher order than that of the numerator polynomial (see, e.g., [114]). The latter particularly requires a strictly proper $G(s)$ to necessarily come with $D = 0$.

1.2 Hardy Space and $\mathcal{H}_\infty$ Norm

The Hardy space $\mathcal{H}_\infty(\mathbb{C}_+)$ on the complex right half-plane $\mathbb{C}_+$ consists of bounded functions on the imaginary axis with analytic continuation into the right half-plane. In the present context, $\mathcal{H}_\infty(\mathbb{C}_+^{l \times m})$ is the set of stable and proper transfer functions $G(s) \in \mathbb{C}^{l \times m}$. The $\mathcal{H}_\infty$ norm $\|\cdot\|_\infty$ of such a transfer function $G(s)$ is defined as the supremum of its maximum singular values $\sigma_{max}\{G(j\omega)\}$ in the frequency domain, that is,

$$\|G(\cdot)\|_\infty \triangleq \operatorname*{ess\,sup}_{\omega} \sigma_{max}\{G(j\omega)\}. \tag{1.10}$$

Recall that given a fixed frequency ω, the matrix $G(j\omega)$ possesses singular values

$$\sigma\{G(j\omega)\} = \lambda^{1/2}\{G^*(j\omega)G(j\omega)\}, \tag{1.11}$$

which are the positive square roots of the eigenvalues $\lambda\{G^*(j\omega)G(j\omega)\}$ of $G^*(j\omega)G(j\omega)$, where $G^*(j\omega)$ is the complex conjugate transpose of $G(j\omega)$.

A useful characterization of the $\mathcal{H}_\infty$ norm is given in terms of a hypothetical input–output experiment. Suppose that an input $u \in \mathcal{L}_2$ is applied to the plant, and consider the output

$$z = G(s)u \in \mathcal{L}_2, \tag{1.12}$$

where $\mathcal{L}_2 \triangleq \mathcal{L}_2(-\infty, \infty)$ is the Hilbert space of square-integrable functions on $(-\infty, \infty)$, and the Laplace transform-based isomorphism of $\mathcal{L}_2$ in the time domain with $\mathcal{L}_2$ in the frequency domain is in liberal use. Then a standard result interprets the $\mathcal{H}_\infty$ norm as the induced norm of the multiplication operator G in (1.12):

$$\|G(\cdot)\|_\infty = \sup_{u \neq 0} \frac{\|z\|_{\mathcal{L}_2}}{\|u\|_{\mathcal{L}_2}} = \sup_{\|u\|_{\mathcal{L}_2}=1} \|z\|_{\mathcal{L}_2}. \tag{1.13}$$

Clearly, the $\mathcal{H}_\infty$ norm, thus interpreted, yields the worst-case gain for sinusoidal inputs $u(\omega)$ at any frequency ω. Indeed, letting $t \to \infty$ and setting $z(\omega)$, the system response, to a persistent sinusoidal input $u(\omega)$, one obtains $z(\omega) = G(j\omega)u(\omega)$. Hence, at a fixed frequency, the gain $\|z(\omega)\|_{\mathcal{L}_2}/\|u(\omega)\|_{\mathcal{L}_2}$ depends on the direction $u(\omega)$; in the worst-case direction, it is given by the maximum singular value:

$$\sigma\{G(j\omega)\} = \sup_{u \neq 0} \frac{\|z\|_{\mathcal{L}_2}}{\|u\|_{\mathcal{L}_2}}. \tag{1.14}$$

Since the gain also depends on the frequency, at the worst-case frequency, it is given by the $\mathcal{H}_\infty$ norm:

$$\|G(\cdot)\|_\infty = \operatorname*{ess\,sup}_{\omega} \sup_{u \neq 0} \frac{\|z\|_{\mathcal{L}_2}}{\|u\|_{\mathcal{L}_2}}. \tag{1.15}$$

If confined to a scalar transfer function $G(s) \in \mathbb{C}$, the $\mathcal{H}_\infty$-norm definition (1.10) is simplified to

$$\|G(\cdot)\|_\infty \triangleq \operatorname*{ess\,sup}_\omega |G(j\omega)|. \qquad (1.16)$$

Thus, the $\mathcal{H}_\infty$ norm of a scalar transfer function $G(s)$ is simply the peaking value of its magnitude. As illustrated below, the computation of the $\mathcal{H}_\infty$ norm (1.16) in this case proves to be rather technical.

Example 1. Consider a stable transfer function

$$G(s) = \frac{1}{s+a}, \quad a > 0,$$

of first order. By definition, its $\mathcal{H}_\infty$ norm is given by

$$\|G(s)\|_\infty = \operatorname*{ess\,sup}_\omega |G(j\omega)| = \operatorname*{ess\,sup}_\omega \left\{ \frac{1}{\sqrt{\omega^2 + a^2}} \right\} = \frac{1}{a}. \qquad (1.17)$$

Hereinafter, routine computations are left to the reader.

Example 2. Consider a stable transfer function

$$G(s) = \frac{\omega_n^2}{s^2 + 2\zeta\omega_n s + \omega_n^2}, \quad \omega_n > 0, \zeta \in \left(0, \frac{1}{\sqrt{2}}\right) \qquad (1.18)$$

of second order. By definition, its $\mathcal{H}_\infty$ norm is given by

$$\|G(s)\|_\infty = \operatorname*{ess\,sup}_\omega |G(j\omega)| = \operatorname*{ess\,sup}_\omega \frac{\omega_n^2}{\sqrt{(\omega_n^2 - \omega^2)^2 + 4(\zeta\omega_n\omega)^2}}. \qquad (1.19)$$

The extremal frequency $\omega^* = \omega_n \sqrt{1 - 2\zeta^2}$ in (1.19) is determined by the relation

$$\left.\frac{d|G(j\omega)|}{d\omega}\right|_{\omega=\omega^*} = 0. \qquad (1.20)$$

Since the second derivative

$$\left.\frac{d^2|G(j\omega)|}{d\omega^2}\right|_{\omega=\omega^*} = -\frac{1}{2}\frac{1-2\zeta^2}{\zeta^3\omega_n^4(1-\zeta^2)^{3/2}} \qquad (1.21)$$

of $|G(j\omega)|$, computed at the extremal frequency ω^*, is negative, the magnitude $|G(j\omega)|$ of the frequency response achieves its maximal value at the extremal frequency. Thus, the $\mathcal{H}_\infty$ norm of the transfer function (1.18) is explicitly computed as

$$\|G(s)\|_\infty = \frac{\omega_n^2}{\sqrt{(\omega_n^2 - {\omega^*}^2)^2 + 4(\zeta\omega_n\omega^*)^2}} = \frac{1}{2\zeta\sqrt{1-\zeta^2}}. \qquad (1.22)$$

1.2 Hardy Space and $\mathcal{H}_\infty$ Norm

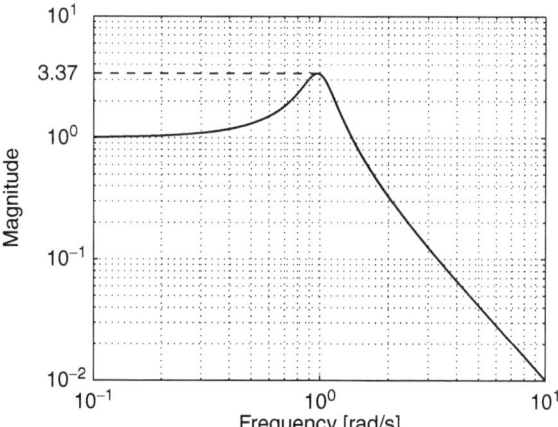

Fig. 1.1 Frequency response (Bode plot) of the transfer function (1.18) specified with $\omega_n = 1$ and $\zeta = 0.15$

The frequency response of the transfer function (1.18), specified with $\omega_n = 1$ and $\zeta = 0.15$, is plotted in Fig. 1.1 in a log-scale where the peak $|G(j\omega^*)| \approx 3.37$ of the plot is achieved at the extremal frequency $\omega^* \approx 0.98$. According to (1.22), it is this peak value that determines the $\mathcal{H}_\infty$ norm of the transfer function (1.18), that is, $\|G(s)\|_\infty \approx 3.37$.

1.2.1 Computing the $\mathcal{H}_\infty$ Norm of Proper Transfer Function Matrices

The straightforward computation of the $\mathcal{H}_\infty$ norm (1.10) of transfer function matrices, while possible in principle, appears to be much lengthier compared to that of scalar transfer functions. In this regard, alternative numerical approaches have been developed and implemented in MATLAB®'s Robust Control toolbox [12, 108].

Such an approach reduces the problem of computing the $\mathcal{H}_\infty$ norm of the transfer function matrix (1.4) of a linear stable system (1.3) to the following optimization problem in $P = P^T > 0$ (see [53] for details):

$$\min \left\{ \gamma > 0 : \begin{bmatrix} A^T P + PA & PB & C^T \\ B^T P & -\gamma I & D^T \\ C & D & -\gamma I \end{bmatrix} < 0 \right\}. \qquad (1.23)$$

The linear matrix inequality (LMI) problem (1.23) is readily solved using MATLAB®. It is worth noticing that γ is uniquely determined by the LMI problem (1.23), whereas P is, generally speaking, nonunique.

1.2.2 Computing the $\mathcal{H}_\infty$ Norm of Strictly Proper Transfer Function Matrices

In order to present another approach to computing the $\mathcal{H}_\infty$ norm of a strictly proper transfer function matrix, consider a particular form

$$G(s) = \left[\begin{array}{c|c} A & B \\ \hline C & 0 \end{array}\right] \qquad (1.24)$$

of (1.4) with $D = 0$ and $A \in \mathbb{R}^{n \times n}$ a Hurwitz matrix, all of whose eigenvalues are in the open left half-plane.

According to [38, Lemma 4], the $\mathcal{H}_\infty$ norm of such a transfer function matrix (1.24) is less than $\gamma > 0$ if and only if the Hamiltonian matrix

$$H = \begin{bmatrix} A & \gamma^{-2} B B^T \\ -C^T C & -A^T \end{bmatrix} \qquad (1.25)$$

has no eigenvalues on the imaginary axis. Due to this property, the $\mathcal{H}_\infty$ norm $\|G\|_\infty$ of a transfer function matrix (1.24) with stable A is iteratively computed as follows. First, a positive number γ is tested if $\|G\|_\infty < \gamma$ by calculating the eigenvalues of (1.25). Then γ is decreased if (1.25) has no eigenvalues on the imaginary axis; it is increased otherwise. By iteration on γ, the $\mathcal{H}_\infty$ norm of the transfer function matrix (1.24) is thus approached.

1.3 Optimal and Suboptimal Synthesis

The $\mathcal{H}_\infty$ (sub)optimal control problem is typically stated for a linear system depicted in Fig. 1.2 with the state $x(t) \in \mathbb{R}^n$, the control input $u(t) \in \mathbb{R}^m$, the unknown disturbance $w(t) \in \mathbb{R}^r$, the output $z(t) \in \mathbb{R}^l$ to be controlled, and the available measurement $y(t) \in \mathbb{R}^p$, imposed on the system. Both the plant G and the controller K in the block diagram of Fig. 1.2 are real-rational and proper.

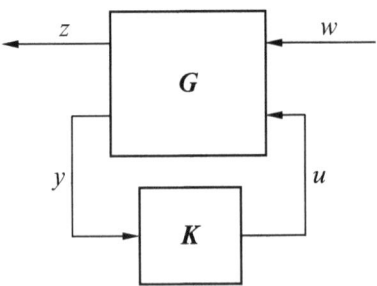

Fig. 1.2 Linear feedback configuration

1.3 Optimal and Suboptimal Synthesis

The control objective is to synthesize an admissible controller K that internally stabilizes the plant while also minimizing (attenuating) the $\mathcal{H}_\infty$ norm of the resulting closed-loop transfer function matrix T_{zw} from w to z. Recall that the closed-loop system is said to be *internally stable* iff it is exponentially stable provided that no disturbance affects the system.

1.3.1 Plant Description and Basic Assumptions

To facilitate exposition, the realization of the transfer matrix G chosen for treatment is given by

$$\left[\begin{array}{c|cc} A & B_1 & B_2 \\ \hline C_1 & 0 & D_{12} \\ C_2 & D_{21} & 0 \end{array}\right], \tag{1.26}$$

which corresponds to the following state-space representation:

$$\begin{aligned} \dot{x} &= Ax + B_1 w + B_2 u, \\ z &= C_1 x + D_{12} u, \\ y &= C_2 x + D_{21} w \end{aligned} \tag{1.27}$$

with matrices $A, B_1, B_2, C_1, C_2, D_{12}, D_{21}$ of appropriate dimensions. The general state-space representation with nonzero feedthrough terms D_{11} and D_{22} can be treated as in [106] by constructing an equivalent problem with $D_{11} = 0$ and $D_{22} = 0$. In addition, the simplifying assumptions

(A1) (A, B_1) is stabilizable and (C_1, A) is detectable,
(A2) (A, B_2) is stabilizable and (C_2, A) is detectable,
(A3) $D_{12}^T C_1 = 0$ and $D_{12}^T D_{12} = I$,
(A4) $B_1 D_{21}^T = 0$ and $D_{21} D_{21}^T = I$,

presented in [38], are made throughout.

Assumption (A1) guarantees that internal stability of G is equivalent to the input–output stability from w to z, whereas Assumption (A2) is necessary and sufficient for G to be internally stabilizable. As shown, for example, in [2], these assumptions can be relaxed since they are ensured by the existence of proper solutions of Riccati equations that appear in solving the $\mathcal{H}_\infty$-control problem in question.

Assumption (A3) deals with orthogonal $C_1 x$ and $D_{12} u$, thereby yielding a nonsingular, normalized penalty on the control u in the $\mathcal{L}_2$ norm of the output $z = C_1 x + D_{12} u$ representing the system performance. As a matter of fact, this means that there is no cross weighting between the control input and the state, and the control weight matrix is the identity. Other nonsingular weights are readily converted to the present problem via a change of coordinates in u. Removing the full-rank condition and dealing with a singular D_{12} have been done in [116] based on the singular control theory. A less direct manner of converting such a singular problem into the standard one has been proposed in [88] based on the Goh integral transformation of the control input.

Assumption (A4) is dual to (A3) and concerns how plant disturbance and sensor noise, forming the exogenous signal w, enter G. Being nonsingular, the sensor noise weighting is normalized, and it is orthogonal to plant disturbance.

1.3.2 Problem Statement

In contrast to the state-space model (1.26) of the plant G, a state-space model of the controller K is not prescribed since it is to be determined from the control objectives. Rather, basic input–output properties have been required of any admissible controller to be real-rational and proper for ensuring that it is a *causal function* of the measured output y (i.e., a function, independent of future values of its argument). In other words, the class of admissible controllers K are those with finite-dimensional linear state-space realizations

$$\begin{aligned} \dot{\xi} &= A_K \xi + B_{1K} y + B_{2K} u, \\ u &= C_K \xi + D_K y, \end{aligned} \tag{1.28}$$

where $\xi \in \mathbb{R}^k$ is the internal state of the controller K, and A_K, B_{1K}, B_{2K}, C_K, and D_K are matrices of appropriate dimensions.

With the feedback configuration of Fig. 1.2 and the assumptions above, the $\mathcal{H}_\infty$ *optimal control problem* is to find an internally stabilizing controller (1.28) that minimizes the $\mathcal{H}_\infty$ norm $\|T_{zw}(G, K)\|_\infty$ of the resulting closed-loop transfer function matrix $T_{zw}(G, K)$ from w to z.

Let γ_{min} be the minimum value of $\|T_{zw}(G, K)\|_\infty$ over all internally stabilizing controllers K. Then $\mathcal{H}_\infty$ *suboptimal control problem* is to find an internally stabilizing controller (1.28) such that

$$\|T_{zw}(G, K)\|_\infty < \gamma \tag{1.29}$$

for an a priori given disturbance attenuation level $\gamma > \gamma_{min}$. Inequality (1.29) admits a transparent representation in terms of the so-called strict bounded-real condition [23, Sect. 2.7.3]:

$$T_{zw}^*(G, K)(s) T_{zw}(G, K)(s) < \gamma I \text{ for all } \mathbf{Re}\, s > 0, \tag{1.30}$$

where $\mathbf{Re}\, s$ stands for the real part of s and I is the identity matrix.

1.3 Optimal and Suboptimal Synthesis

1.3.3 Problem Solution

The necessary and sufficient conditions, established in the seminal work [38] for the above $\mathcal{H}_\infty$ suboptimal control problem to have a solution with a disturbance attenuation level $\gamma > 0$, are summarized below.

(C1) The equation

$$PA + A^T P + C_1^T C_1 + P \left[\frac{1}{\gamma^2} B_1 B_1^T - B_2 B_2^T \right] P = 0 \qquad (1.31)$$

possesses a positive semidefinite symmetric solution P such that the matrix $A - (B_2 B_2^T - \gamma^{-2} B_1 B_1^T) P$ has all eigenvalues with negative real part.

(C2) The equation

$$AR + RA^T + B_1 B_1^T + R \left[\frac{1}{\gamma^2} C_1^T C_1 - C_2^T C_2 \right] R = 0 \qquad (1.32)$$

possesses a positive semidefinite symmetric solution R such that the matrix $A - R(C_2^T C_2 - \gamma^{-2} C_1^T C_1)$ has all eigenvalues with negative real part.

(C3) The matrix PR has spectral radius $\rho(PR)$ strictly less than γ; that is,

$$\rho(PR) < \gamma. \qquad (1.33)$$

Following [114, Sect. 9.3.4], the suboptimality tests are formulated in terms of the solvability of the AREs (1.31), (1.32) and the coupling condition (1.33) on their solutions.

Theorem 1. *The $\mathcal{H}_\infty$ suboptimal control problem possesses a solution for the plant model (1.26) with Assumptions (A1)–(A4) and a disturbance attenuation level $\gamma > 0$ if and only if Conditions (C1)–(C3) are satisfied. Once these conditions hold, the so-called central controller*

$$K_{sub} = \left[\begin{array}{c|c} \hat{A} & -ZL \\ \hline F & 0 \end{array} \right] \qquad (1.34)$$

yields a solution to the problem in question provided that

$$\begin{aligned}
\hat{A} &= A + \gamma^{-2} B_1 B_1^T P + B_2 F + ZLC_2, \\
F &= -B_2^T P, \\
L &= -RC_2^T, \\
Z &= (I - \gamma^{-2} RP)^{-1}.
\end{aligned} \qquad (1.35)$$

An efficient test is thus provided by Theorem 1 for each value of γ to verify, using MATLAB®, if it is less or greater than the minimal disturbance attenuation level γ_{min}. If we reduce γ iteratively, an *optimal solution* is approached with the corresponding central controller (1.34).

If we want to reveal a useful separation structure of the central controller (1.34), it suffices to represent it in the state-space form

$$\dot{\xi} = A\xi + B_1\hat{w}_{worst} + B_2 u + ZL(C_2\xi - y), \qquad (1.36)$$
$$u = F\xi, \quad \hat{w}_{worst} = \gamma^{-2} B_1^T P\xi,$$

where notation (1.35) has been taken into account. In the state-space representation (1.36), the central controller is clearly separated into a state estimator (observer), where ZL is the optimal filter gain for estimating the optimal state feedback Fx in the presence of the worst-case input $w_{worst} = \gamma^{-2} B_1^T Px$. In these separation terms, Condition ($C1$) is necessary and sufficient for the $\mathcal{H}_\infty$ suboptimal control problem to have a solution in a particular case of the state-feedback design where the full information on the state is available. In turn, Condition ($C2$) is necessary and sufficient for the $\mathcal{H}_\infty$ suboptimal control problem to have a solution in a dual case of the state estimation where the output injection has full access to the state of the estimator. Condition ($C3$) couples elements from ($C1$) and ($C2$).

It is clear that the latter condition is hardly extendible to nonautonomous systems, for instance, and an alternative coupling condition, not dealing with a spectral radius, would be of interest both in itself and due to potential extensions to a wider class of uncertain systems. Such an alternative occurs if the Riccati equation (1.32) is represented in terms of the symmetric matrix $Q = R(I - \gamma^{-2}PR)^{-1}$, where P and R are governed by (1.31) and (1.32), respectively. Since the matrix $I - \gamma^{-2}PR$ is invertible and positive semidefinite provided that Conditions ($C1$)–($C3$) hold, the resulting Riccati equation on $Q = Q^T \geq 0$ is obtained in the form

$$A_1 Q + Q A_1^T + B_1 B_1^T + Q \left[\frac{1}{\gamma^2} P B_2 B_2^T P - C_2^T C_2 \right] Q = 0, \qquad (1.37)$$

specified with

$$A_1 = A + \frac{1}{\gamma^2} B_1 B_1^T P \qquad (1.38)$$

and coupled with an appropriate solution P of (1.31). It is known from [38] that Conditions ($C1$)–($C3$) are equivalent to ($C1$) coupled with

($C4$) Equation (1.37) specified with (1.38) possesses a positive semidefinite symmetric solution Q such that the matrix $A_1 - Q(C_2^T C_2 - \gamma^{-2} P B_2 B_2^T P)$ has all eigenvalues with negative real part.

A suboptimality test, alternative to Theorem 1, is thus reworked.

1.3 Optimal and Suboptimal Synthesis

Theorem 2. *The $\mathcal{H}_\infty$ suboptimal control problem possesses a solution for the plant model (1.26) with Assumptions (A1)–(A4) and a disturbance attenuation level $\gamma > 0$ if and only if Conditions (C1) and (C4), coupled together, are satisfied. Once these conditions hold, the central controller*

$$\dot{\xi} = A\xi + \left(\frac{1}{\gamma^2} B_1 B_1^T - B_2 B_2^T\right) P\xi + QC_2^T(y - C_2\xi), \quad (1.39)$$
$$u = -B_2^T P\xi$$

yields a solution to the problem in question.

Remarkably, Theorem 2 admits a rather straightforward extension to nonautonomous systems that will be presented in the next section.

1.3.4 LMI-Based Solution

Along with the AREs (1.31) and (1.32), consider their inequality counterparts

$$\mathcal{P}A + A^T\mathcal{P} + C_1^T C_1 + \mathcal{P}\left[\frac{1}{\gamma^2} B_1 B_1^T - B_2 B_2^T\right]\mathcal{P} < 0, \quad (1.40)$$

$$A_1 \mathcal{Q} + \mathcal{Q}A_1^T + B_1 B_1^T + \mathcal{Q}\left[\frac{1}{\gamma^2}\mathcal{P} B_2 B_2^T \mathcal{P} - C_2^T C_2\right]\mathcal{Q} < 0, \quad (1.41)$$

with the focus on *the feasibility problem* of seeking some positive definite matrices $\mathcal{P}, \mathcal{Q}$, which may not be unique, such that (1.40) and (1.41) hold.

For the purpose of representing the above quadratic inequalities in the form of LMIs, solvable with MATLAB®, let's first multiply either side of (1.40) and (1.41) by $\Pi = \mathcal{P}^{-1} > 0$ and $\Theta = \mathcal{Q}^{-1} > 0$, respectively. The resulting inequalities,

$$A\Pi + \Pi A^T + \Pi C_1^T C_1 \Pi + \frac{1}{\gamma^2} B_1 B_1^T - B_2 B_2^T < 0, \quad (1.42)$$

$$\Theta A_1 + A_1^T \Theta + \Theta B_1 B_1^T \Theta + \frac{1}{\gamma^2} \Pi^{-1} B_2 B_2^T \Pi^{-1} - C_2^T C_2 < 0, \quad (1.43)$$

although still quadratic, are then readily converted, by applying the Schur complements' formula, into the LMIs

$$\Phi(\Pi) = \begin{bmatrix} A\Pi + \Pi A^T + \frac{1}{\gamma^2} B_1 B_1^T - B_2 B_2^T & \Pi C_1^T \\ C_1 \Pi & -I \end{bmatrix} < 0, \quad (1.44)$$

$$\Psi(\Theta) = \begin{bmatrix} \Theta A_1 + A_1^T \Theta + \frac{1}{\gamma^2} \Pi^{-1} B_2 B_2^T \Pi^{-1} - C_2^T C_2 & \Theta B_1 \\ B_1^T \Theta & -I \end{bmatrix} < 0, \quad (1.45)$$

where the former is an independent LMI in $\Pi > 0$, whereas the latter is an LMI in $\Theta > 0$, coupled with an a priori computed feasible solution $\Pi > 0$ of (1.44). To reflect the fact that $\Pi > 0$ stands in (1.45) for a feasible solution of (1.44), computed a priori, the argument of the matrix $\Psi(\Theta)$ is confined to the variable Θ to be determined as a feasible solution of (1.45).

Furthermore, the LMIs (1.44), (1.45) and the requirements $\Pi > 0$, $\Theta > 0$ can be combined into a couple of single LMIs as

$$\begin{bmatrix} \Phi(\Pi) & 0 \\ 0 & -\Pi \end{bmatrix} < 0 \quad \text{and} \quad \begin{bmatrix} \Psi(\Theta) & 0 \\ 0 & -\Theta \end{bmatrix} < 0, \qquad (1.46)$$

thereby admitting the use of the MATLAB® LMI toolbox [12, 108] for solving the feasibility problem in question.

Summarizing, we obtain the following result.

Theorem 3. *The ARIs (1.40) and (1.41) possess positive definite solutions $\mathcal{P}$ and $\mathcal{Q}$ if and only if $\Pi = \mathcal{P}^{-1}$ and $\Theta = \mathcal{Q}^{-1} > 0$ are feasible solutions of the corresponding LMIs in (1.46).*

Proof. Necessity: Let $\mathcal{P}$ and $\mathcal{Q}$ be positive definite solutions of (1.40) and (1.41). Multiplication of either side of (1.40) and (1.41) by the nonsingular matrices $\Pi = \mathcal{P}^{-1} > 0$ and $\Theta = \mathcal{Q}^{-1} > 0$ does not change the definiteness of the resulting expressions (1.42) and (1.43), which are therefore satisfied with the positive definite matrices $\Pi = \mathcal{P}^{-1}$ and $\Theta = \mathcal{Q}^{-1}$. Moreover, due to Schur complements, a matrix

$$\Phi = \begin{bmatrix} \Phi_{11} & \Phi_{12} \\ \Phi_{12}^T & \Phi_{22} \end{bmatrix}, \qquad (1.47)$$

for example, specified in accordance with (1.44), is negative definite if and only if

$$\Phi_{22} < 0 \quad \text{and} \quad \Phi_{11} - \Phi_{12}\Phi_{22}^{-1}\Phi_{12}^T < 0. \qquad (1.48)$$

The equivalence of the feasibility of (1.44) to that of (1.42) is thus validated. The equivalence of the feasibility of (1.45) to that of (1.43) is then validated in a similar manner. For us to complete the proof of the necessity part, it remains to note that the LMIs (1.44), (1.45) under the requirements $\Pi > 0$, $\Theta > 0$ are equivalent to (1.46).

Sufficiency: As has been demonstrated, feasible solutions Π and Θ of (1.46) constitute positive definite solutions of (1.42) and (1.43). By multiplying (1.42) and (1.43) by $\mathcal{P} = \Pi^{-1}$ and $\mathcal{Q} = \Theta^{-1}$, we show that the resulting inequalities (1.40) and (1.41) hold with $\mathcal{P} = \Pi^{-1} > 0$ and $\mathcal{Q} = \Theta^{-1} > 0$. This completes the proof of Theorem 3.

As is clear from the direct state-space approach [97], the ARIs are also capable of forming a basis for studying the $\mathcal{H}_\infty$-control problem. While dealing with such inequalities, Conditions $(C1)$ and $(C4)$ are modified to

1.3 Optimal and Suboptimal Synthesis

($C1$) Inequality (1.40) possesses a positive definite symmetric solution $\mathcal{P}$.
($C4$) Inequality (1.41) specified with (1.38) possesses a positive definite symmetric solution $\mathcal{Q}$.

to remain necessary and sufficient for a solution of the $\mathcal{H}_\infty$ suboptimal control problem to exist.

Theorem 4. *The $\mathcal{H}_\infty$ suboptimal control problem possesses a solution for the plant model (1.27) with Assumptions (A1)–(A4) and a disturbance attenuation level $\gamma > 0$ if and only if Conditions ($C1$) and ($C4$), coupled together, are satisfied. Once these conditions hold, the controller*

$$\dot{\xi} = A\xi + \left(\frac{1}{\gamma^2} B_1 B_1^T - B_2 B_2^T\right) \mathcal{P}\xi + \mathcal{Q}C_2^T(y - C_2\xi),$$

$$u = -B_2^T \mathcal{P}\xi \tag{1.49}$$

yields a solution to the problem in question.

Proof. Necessity: Let the $\mathcal{H}_\infty$ suboptimal control problem possess a solution. Then Theorem 2 ensures that Conditions ($C1$) and ($C4$) hold. By applying the strict bounded real lemma, such as that from [5] (see also the ongoing Lemma 7 for its periodic version that represents the time-invariant case, in particular), it follows that for sufficiently small $\varepsilon > 0$ and the identity matrix $I \in \mathbb{R}^{n \times n}$, there exist positive definite solutions P_ε and Q_ε of the perturbed Riccati equations

$$P_\varepsilon A + A^T P_\varepsilon + C_1^T C_1 + P_\varepsilon \left[\frac{1}{\gamma^2} B_1 B_1^T - B_2 B_2^T\right] P_\varepsilon = -\varepsilon I, \tag{1.50}$$

$$A_1 Q_\varepsilon + Q_\varepsilon A_1^T + B_1 B_1^T + Q_\varepsilon \left[\frac{1}{\gamma^2} P_\varepsilon B_2 B_2^T P_\varepsilon - C_2^T C_2\right] Q_\varepsilon = -\varepsilon I. \tag{1.51}$$

As a matter of fact, the positive definite matrices $\mathcal{P} = P_\varepsilon$ and $\mathcal{Q} = Q_\varepsilon$ represent feasible solutions of the ARIs (1.42) and (1.43). The validity of Conditions ($C1$) and ($C4$) is thus established.

Sufficiency: Provided that Conditions ($C1$) and ($C4$) are satisfied, let's first introduce the functions

$$V(x) = x^T \mathcal{P} x, \tag{1.52}$$

where $\mathcal{P}$ is a solution of the ARI (1.42), and

$$H(x, p^T, w, u) = p^T[Ax + B_1 w + B_2 u] + \|C_1 x + D_{12} u\|^2 - \gamma^2 \|w\|^2. \tag{1.53}$$

These functions come with the following features.

1. Function (1.52) is positive definite and radially unbounded, and its time derivative, computed along the trajectories of (1.27), is given by

$$\dot{V}(x(t)) = x^T(t)[\mathcal{P}A + A^T\mathcal{P}]x(t) + 2x^T\mathcal{P}B_1 w + 2x^T\mathcal{P}B_2 u. \quad (1.54)$$

2. Function (1.53) is quadratic in (w, u); under Assumption (A3), it is representable in the form

$$H(x, p^T, w, u) = H(x, p^T, \alpha_1, \alpha_2) - \gamma^2 \|w - \alpha_1\|^2 + \|u - \alpha_2\|^2, \quad (1.55)$$

with

$$\alpha_1(p^T) = \frac{1}{2}\gamma^{-2} B_1^T p, \quad \alpha_2(p^T) = -\frac{1}{2} B_2^T p \quad (1.56)$$

being unique extremal points of the function H where its partial derivatives in w and u are nullified; that is,

$$\frac{\partial H(x, p^T, w, u)}{\partial w} = 0 \quad \text{and} \quad \frac{\partial H(x, p^T, w, u)}{\partial u} = 0 \quad \text{at } (w, u) = (\alpha_1, \alpha_2).$$

3. The equality

$$H(x, V_x, \alpha_1(V_x), \alpha_2(V_x)) = \mathcal{P}A + A^T\mathcal{P} + C_1^T C_1 + \mathcal{P}\left[\frac{1}{\gamma^2} B_1 B_1^T - B_2 B_2^T\right]\mathcal{P} \quad (1.57)$$

is straightforwardly verified with $V(x)$, governed by (1.52), $V_x = \partial V/\partial x$, and α_1, α_2, given by (1.56).

4. Due to (1.56) and (1.57), the Hamilton–Jacobi–Isaacs partial differential inequality

$$H(x, V_x, \alpha_1(V_x), \alpha_2(V_x)) < 0 \quad (1.58)$$

becomes feasible with $V(x)$, governed by (1.52).

5. Relations (1.55), (1.56), and (1.58) result in

$$H(x, V_x, w, u) < -\gamma^2 \|w - \alpha_1(V_x)\|^2 + \|u - \alpha_2(V_x)\|^2, \quad (1.59)$$

and due to (1.52)–(1.54), coupled with Assumption (A3), it follows that

$$\dot{V}(x(t)) = H(x, V_x, w, u) - \|C_1 x + D_{12} u\|^2 + \gamma^2 \|w\|^2$$
$$< -\gamma^2 \|w - \alpha_1(V_x)\|^2 + \|u - \alpha_2(V_x)\|^2 - \|C_1 x\|^2 - \|u\|^2 + \gamma^2 \|w\|^2. \quad (1.60)$$

Remark 1. In a particular case of the full state measurements $y = x$ with $C_2 = I$ and $D_{21} = 0$, the above features ensure that the state feedback $u = \alpha_2(V_x) = -B_2^T \mathcal{P} x$ solves the $\mathcal{H}_\infty$ suboptimal control problem in question. Indeed, the internal

1.3 Optimal and Suboptimal Synthesis

dynamics of the corresponding closed-loop system (1.27) are exponentially stable because relation (1.53) determines a Lyapunov function of the closed-loop system whose temporal derivative (1.54), computed along these unperturbed dynamics, is simplified to

$$\dot{V}(x(t)) = x^T(t)[\mathcal{P}A + A^T\mathcal{P} - 2\mathcal{P}B_2 B_2^T \mathcal{P}]x(t), \qquad (1.61)$$

where, when one accounts for (1.40), the quadratic form on the right-hand side is readily shown to be negative definite:

$$\mathcal{P}A + A^T\mathcal{P} - 2\mathcal{P}B_2 B_2^T \mathcal{P}$$

$$< \mathcal{P}A + A^T\mathcal{P} + C_1^T C_1 + \mathcal{P}\left[\frac{1}{\gamma^2}B_1 B_1^T - B_2 B_2^T\right]\mathcal{P} < 0. \qquad (1.62)$$

In addition, given a trajectory $x(t)$ of the closed-loop system (1.27), driven by the state feedback $u = \alpha_2(V_x)$, relations (1.52)–(1.58) result in

$$\dot{V}(x(t)) = H(x, V_x, w, \alpha_2(V_x)) - \|C_1 x + D_{12}\alpha_2(V_x)\|^2 + \gamma^2 \|w\|^2$$
$$= H(x, V_x, \alpha_1, \alpha_2) - \gamma^2 \|w - \alpha_1\|^2 - \|z\|^2 + \gamma^2 \|w\|^2$$
$$< -\gamma^2 \|w - \alpha_1\|^2 - \|z\|^2 + \gamma^2 \|w\|^2, \qquad (1.63)$$

thereby ensuring that

$$V(x(t)) - V(x(0)) <$$
$$-\int_0^t [\gamma^2 \|w(\tau) - \gamma^{-2} B_1^T \mathcal{P} x(\tau)\|^2 + \|z(\tau)\|^2 - \gamma^2 \|w(\tau)\|^2] d\tau. \qquad (1.64)$$

Since the function $V(x)$ is positive definite, it follows that

$$\int_0^\infty \|z(\tau)\|^2 d\tau < \gamma^2 \int_0^\infty \|w(\tau)\|^2 d\tau \qquad (1.65)$$

for any $w \in \mathcal{L}_2$ and $x(t)$, initialized at $x(0) = 0$. Thus, by the $\mathcal{H}_\infty$-induced norm definition (1.13), the integral inequality (1.65) verifies that in the full state measurement case, the $\mathcal{H}_\infty$ norm of the closed-loop system is indeed less than γ.

Next, let's introduce the functions

$$W(x, \xi) = \gamma^2 (x - \xi)^T Q^{-1}(x - \xi) \qquad (1.66)$$

and

$$H_e(x, \xi, r) = 2\gamma^2 (x - \xi)^T \{Q^{-1}[Ax + B_1 w_{worst}(x) + B_2 u_{opt}(\xi) + B_1 r]$$
$$- Q^{-1}[A\xi + B_1 w_{worst}(\xi) + B_2 u_{opt}(\xi)] + C_2^T (C_2 x - C_2 \xi) + C_2^T D_{21} r\}$$
$$+ [u_{opt}(\xi) - u_{opt}(x)]^T [u_{opt}(\xi) - u_{opt}(x)] - \gamma^2 r^T r, \qquad (1.67)$$

where Q is a solution of the ARI (1.43), and

$$w_{worst}(x) = \gamma^{-2} B_1^T \mathcal{P} x, \quad u_{opt}(\xi) = -B_2^T \mathcal{P} \xi. \tag{1.68}$$

Attractive features of these functions, listed below, are similar to those of (1.52) and (1.53).

1. Function (1.66) is positive semidefinite, whereas $W(0, \xi)$ is positive definite and radially unbounded, and if we take into account that $D_{21} w_{worst}(x) = 0$ by Assumption (A4), the time derivative, computed along the trajectories of (1.27), (1.49), is given by

$$\dot{W}(x(t), \xi(_2 u]t)) = 2\gamma^2 (x - \xi)^T \{ Q^{-1}[Ax + B_1 w_{worst}(x) \\ + B_1(w - w_{worst}(x)) + B_2] - Q^{-1}[A\xi + B_1 w_{worst}(\xi) \\ + B_2 u_{opt}(\xi)] + C_2^T (C_2 x + D_{21}(w - w_{worst}(x)) - C_2 \xi) \}. \tag{1.69}$$

2. Function (1.67) is quadratic in r and is representable in the form

$$H_e(x, \xi, r) = H_e(x, \xi, \phi(x, \xi)) - \gamma^2 \| r - \phi(x, \xi) \|^2 \tag{1.70}$$

with

$$\phi(x, \xi) = [B_1^T Q^{-1} - D_{21}^T C_2](x - \xi), \tag{1.71}$$

as a unique extremal point of the function H_e, where its partial derivative in r is nullified, that is, $\partial H_e(x, \xi, r) / \partial r = 0$.

3. Under Assumption (A4), the relation

$$H_e(x, \xi, \phi(x, \xi)) = Q^{-1} A_1 + A_1^T Q^{-1} + Q^{-1} B_1 B_1^T Q^{-1} \\ + \frac{1}{\gamma^2} \mathcal{P} B_2 B_2^T \mathcal{P} - C_2^T C_2 < 0 \tag{1.72}$$

is straightforwardly verified with A_1, specified by (1.38), whereas the latter inequality results from (1.41).

4. Taking into account (1.70) and (1.72), the inequality

$$H_e(x, \xi, r) < -\gamma^2 \| r - \phi(x, \xi) \|^2 \tag{1.73}$$

holds.

5. Relations (1.66)–(1.69) result in

$$\dot{W}(x(t), \xi(t)) = H_e(x, \xi, w - w_{worst}(x)) \\ - \| u_{opt}(\xi) - u_{opt}(x) \|^2 + \gamma^2 \| w - w_{worst}(x) \|^2, \tag{1.74}$$

1.3 Optimal and Suboptimal Synthesis

and due to (1.73), it follows that

$$\dot{W}(x(t), \xi(t)) < -\gamma^2 \|w - w_{worst}(x) - \phi(x, \xi)\|^2$$
$$- \|u_{opt}(\xi) - u_{opt}(x)\|^2 + \gamma^2 \|w - w_{worst}(x)\|^2. \quad (1.75)$$

Finally, let's consider the function

$$U(x, \xi) = V(x) + W(x, \xi), \quad (1.76)$$

which is positive definite by virtue of the features of the functions V and W listed above. Taking into account that $\alpha_1(V_x) = w_{worst}(x)$ and $\alpha_2(V_x) = u_{opt}(x)$ by virtue of (1.52), (1.56), and (1.68), differentiating function (1.76) on the trajectories of the closed-loop system (1.27), (1.49) in accordance with (1.60) and (1.75) leads to

$$\dot{U}(x(t), \xi(t)) < -\|C_1 x\|^2 - \|u\|^2 + \gamma^2 \|w\|^2$$
$$- \gamma^2 \|w - w_{worst}(x) - \phi(x, \xi)\|^2. \quad (1.77)$$

Setting $w = 0$ in (1.77) ensures that the time derivative of the positive definite function (1.76) is negative definite along the trajectories of the disturbance-free system (1.27), (1.49), thereby establishing the internal asymptotic stability of the system in question.

In turn, integrating (1.77) yields

$$\int_0^t \|z(\tau)\|^2 d\tau < \int_0^t \|w(\tau)\|^2 d\tau + U(x(0), \xi(0)) - U(x(0), \xi(0)), \quad (1.78)$$

where the relation $\|z\|^2 = \|C_1 x\|^2 + \|u\|^2$, resulting from Assumption ($A3$), has been utilized. Since the function $U(x, \xi)$ is positive definite, it follows that the integral inequality (1.65) holds for any $w \in \mathcal{L}_2$ and $x(t), \xi(t)$, initialized at $x(0) = 0, \xi(0) = 0$. Thus, by the $\mathcal{H}_\infty$-induced norm definition (1.13), this verifies (cf. Remark 1 for the full state measurement case) that the $\mathcal{H}_\infty$ norm of the closed-loop system (1.27), (1.49) is less than γ. This completes the proof of Theorem 4.

Chapter 2
The LMI Approach in an Infinite-Dimensional Setting

Extended via the Lyapunov–Krasovskii method to linear time-delay systems (LTDS), the LMI approach has long been recognized as a powerful analysis tool of such systems. In the present chapter, this approach is further extended to the stability analysis of LTDSs evolving in a Hilbert space. The operator acting on the delayed state is supposed to be bounded. The system delay is unknown and time-varying, with an a priori given upper bound on the delay. Following [50], sufficient exponential stability conditions are derived in the form of linear operator inequalities, where the decision variables are operators in the Hilbert space. When applied to a heat equation and to a wave equation, these conditions are reduced to standard LMIs.

2.1 Historical Remarks and Notation

Time delay naturally appears in many control systems, and it is frequently a source of instability [59]. In the case of distributed parameter systems (DPS)s, even arbitrarily small delays in the feedback may destabilize the system [37, 79, 84, 101]. The stability issue of systems with delay is, therefore, of theoretical and practical value.

During the past decade, a considerable amount of attention has been paid to the stability of ODEs with uncertain constant or time-varying delays (see, e.g., [58, 73, 86, 103]). Special forms of Lyapunov–Krasovskii functionals have been used to derive simple finite-dimensional conditions in terms of LMIs [23]. These conditions are either delay-independent or delay-dependent.

The stability analysis of PDEs with delay is essentially more complicated. There are only a few works on Lyapunov-based technique for PDEs with delay. The second Lyapunov method was extended to abstract nonlinear time-delay systems in the Banach spaces in [126]. Stability and instability conditions for delay wave equations were established in [85]. Stability conditions and exponential bounds were derived in [127, 128] for some scalar heat equations and wave equations with constant delays and with Dirichlet boundary conditions. In [50], the exponential stability of general DPSs was studied within the framework of LTDSs evolving in a Hilbert space.

It is the latter framework that is adopted in the present chapter. Provided the system delay is unknown and time-varying, sufficient *delay-dependent* exponential stability conditions are derived in the form of LOIs, where the decision variables are operators in the Hilbert space. Although available methods for solving LOIs are confined to finite-dimensional approximations [65], however, if exemplified with test heat and wave equations, the derived conditions prove to be reduced to standard finite-dimensional LMIs that are well known to guarantee the exponential stability of the first-order and, respectively, second-order delay-differential equations, describing the first modal dynamics of these test PDEs. Surprisingly, such a *reduction of infinite-dimensional LOIs to finite-dimensional LMIs is tight* in the sense that the stability of the latter delay-differential equations is necessary and sufficient for the stability of the PDEs in question, a feature presently not recognized for ODEs with delay.

The specific notation in the infinite-dimensional setting, used throughout, is fairly standard. The superscript "T" stands for matrix transposition, $\mathbb{R}^n$ denotes the n-dimensional Euclidean space with the norm $|\cdot|$, and $\mathbb{R}^{n \times m}$ is the set of all $n \times m$ real matrices. The notation $P > 0$, for $P \in \mathbb{R}^{n \times n}$, means that P is symmetric and positive definite, whereas $\lambda_{min}(P)$ [$\lambda_{max}(P)$] denotes its minimum (maximum) eigenvalue.

Let $\mathcal{H}$ be a Hilbert space equipped with the inner product $\langle \cdot, \cdot \rangle$ and the corresponding norm $|\cdot|$. Denote by $\mathcal{L}(\mathcal{H})$ bounded linear operators from $\mathcal{H}$ to $\mathcal{H}$. Given a linear operator $P : \mathcal{H} \to \mathcal{H}$ with a dense domain $\mathcal{D}(P) \subset \mathcal{H}$, the notation P^* stands for the adjoint operator. Such an operator P is strictly positive definite, that is, $P > 0$ iff it is self-adjoint in the sense that $P = P^*$ and there exists a constant $\beta > 0$ such that $\langle x, Px \rangle \geq \beta \langle x, x \rangle$ and for all $x \in \mathcal{D}(P)$, whereas $P \geq 0$ means that P is self-adjoint and nonnegative definite, that is, $\langle x, Px \rangle \geq 0$ for all $x \in \mathcal{D}(P)$.

If an infinitesimal operator A generates a strongly continuous semigroup $T(t)$ on the Hilbert space $\mathcal{H}$ (see, e.g., [35] for details), the domain of the operator A forms another Hilbert space $\mathcal{D}(A)$ with the graph inner product $(\cdot, \cdot)_{\mathcal{D}(A)}$ defined as follows: $(x, y)_{\mathcal{D}(A)} = \langle x, y \rangle + \langle Ax, Ay \rangle$, $x, y \in \mathcal{D}(A)$. Moreover, the induced norm $\|T(t)\|$ of the semigroup $T(t)$ satisfies the inequality $\|T(t)\| \leq \kappa e^{\sigma t}$ everywhere with some constant $\kappa > 0$ and growth bound σ.

The space of the continuous $\mathcal{H}$-valued functions $x : [a, b] \to \mathcal{H}$ with the induced norm $\|x\|_{C([a,b],\mathcal{H})} = \max_{s \in [a,b]} |x(s)|$ is denoted by $C([a, b], \mathcal{H})$. The space of the continuously differentiable $\mathcal{H}$-valued functions $x : [a, b] \to \mathcal{H}$, with the induced norm $\|x\|_{C^1([a,b],\mathcal{H})} = \max(\|x\|_{C([a,b],\mathcal{H})}, \|\dot{x}\|_{C([a,b],\mathcal{H})})$, is denoted by $C^1([a, b], \mathcal{H})$.

$L_2(a, b; \mathcal{H})$ is the Hilbert space of square-integrable $\mathcal{H}$-valued functions on (a, b) with the corresponding norm; $W^{l,2}([a, b], R)$ is the Sobolev space of absolutely continuous scalar functions on $[a, b]$ with square-integrable derivatives of the order $l \geq 1$.

The notation $x^t = x(t + \theta) \in L_2(-h, 0; \mathcal{H})$ stands for $x(\cdot) \in L_2(a, b; \mathcal{H})$ and $t \in [a + h, b]$.

2.2 The Lyapunov–Krasovskii Method

For later use, the following instrumental results are extracted from the literature.

Lemma 1 (Wirtinger's inequality and its generalization [127]). *Let* $u \in W^{1,2}([a,b], R)$ *be a scalar function with* $u(a) = u(b) = 0$. *Then*

$$\int_a^b u^2(\xi)d\xi \leq \frac{(b-a)^2}{\pi^2} \int_a^b (u'(\xi))^2 d\xi, \qquad (2.1)$$

$$max_{\xi \in [a,b]} z^2(\xi) \leq (b-a) \int_a^b (z'(\xi))^2 d\xi. \qquad (2.2)$$

If, additionally, $u \in W^{2,2}([a,b], R)$, *then*

$$\int_a^b (u'(\xi))^2 d\xi \leq \frac{(b-a)^2}{\pi^2} \int_a^b (u''(\xi))^2 d\xi. \qquad (2.3)$$

Lemma 2 (Jensen's inequality [58]). *Let* $\mathcal{H}$ *be a Hilbert space with the inner product* $\langle \cdot, \cdot \rangle$. *For any linear bounded operator* $R : \mathcal{H} \to \mathcal{H}$, $R > 0$, *scalar* $l > 0$, *and* $x \in L_2([a,b], \mathcal{H})$, *the following holds:*

$$l \int_0^l \langle x(s), Rx(s) \rangle ds \geq \left\langle \int_0^l x(s)ds, R \int_0^l x(s)ds \right\rangle. \qquad (2.4)$$

Next, consider a linear infinite-dimensional system

$$\dot{x}(t) = Ax(t) + A_1 x(t - \tau(t)), \quad t \geq t_0, \qquad (2.5)$$

evolving in a Hilbert space $\mathcal{H}$, where $x(t) \in \mathcal{H}$ is the instantaneous state of the system. Let the following assumptions be satisfied:

A1 The operator A generates a strongly continuous semigroup $T(t)$ and the domain $\mathcal{D}(A)$ of the operator A is dense in $\mathcal{H}$.
A2 The linear operator A_1 is bounded in $\mathcal{H}$.
A3 The function $\tau(t)$ is piecewise-continuous of class C^1 on the closure of each continuity subinterval and it satisfies

$$\inf_t \tau(t) \geq 0, \ \sup_t \tau(t) \leq h \qquad (2.6)$$

with some constant $h > 0$ for all $t \geq t_0$.

Let the initial conditions

$$x^{t_0} = \varphi(\theta), \ \theta \in [-h, 0], \ \phi \in W, \tag{2.7}$$

be given in the space

$$W = C([-h, 0], \mathcal{D}(A)) \cap C^1([-h, 0], \mathcal{H}). \tag{2.8}$$

Recall that a function $x(t) \in C([t_0 - h, t_0 + \eta], \mathcal{D}(A))$ is said to be a solution of the initial-value problem (2.5), (2.7) on $[t_0 - h, t_0 + \eta]$ if $x(t)$ is initialized with (2.7), it is absolutely continuous for $t \in [t_0, t_0 + \eta]$, and it satisfies (2.5) for almost all $t \in [t_0, t_0 + \eta]$.

The initial-value problem (2.5), (2.7) turns out to be well posed on the semiinfinite time interval $[t_0, \infty)$ and due to Lemma 3 given below, its solutions can be found as mild solutions, namely, as those of the integral equation

$$\begin{aligned} x(t) &= T(t - t_0)x(t_0) \\ &+ \int_{t_0}^{t} T(t - s)A_1 x(s - \tau(s))ds, \ t \geq t_0. \end{aligned} \tag{2.9}$$

The following result is in order.

Lemma 3 ([50]). *Under A1–A3, there exists a unique solution of the initial-value problem (2.5), (2.7) on $[t_0, \infty)$. This solution is also a unique solution of the integral initial-value problem (2.7), (2.9).*

The present aim is to derive exponential stability criteria for linear time-delay systems (2.5), (2.6), thus defined. The stability concept under study is based on the initial data norm

$$\|\phi\|_W = \sqrt{|A\phi(0)|^2 + \|\phi\|^2_{C^1([-h,0],\mathcal{H})}} \tag{2.10}$$

in space (2.8). Suppose $x(t, t_0, \phi)$ denotes a solution of (2.5), (2.7) at a time instant $t \geq t_0$.

Definition 1. System (2.5) is said to be exponentially stable with a decay rate $\delta > 0$ if there exists a constant $K \geq 1$ such that the following exponential estimate holds:

$$|x(t, t_0, \phi)|^2 \leq K e^{-2\delta(t - t_0)} \|\phi\|_W^2 \quad \forall t \geq t_0. \tag{2.11}$$

Consider Lyapunov–Krasovskii functionals, which depend on x and $\dot{x}$ [73]. Given a continuous functional $V : \mathbb{R} \times W \times C([-h, 0], \mathcal{H}) \to \mathbb{R}$, its upper right-hand derivative along solutions $x^t(t_0, \phi), \ t \geq t_0$ of (2.5), (2.7) is defined as follows:

$$\begin{aligned} \dot{V}(t, \phi, \dot{\phi}) &= \limsup_{s \to 0+} \tfrac{1}{s}[V(t+s, x^{t+s}(t, \phi), \dot{x}^{t+s}(t, \phi)) \\ &\quad - V(t, \phi, \dot{\phi})]. \end{aligned}$$

Lemma 4. *Let A1–A3 be in force and let there exist positive numbers δ, β, γ and a continuous functional* such that the function $\bar{V}(t) = V(t, x^t, \dot{x}^t)$ is absolutely continuous for x^t, satisfying (2.5), and

$$\beta|\phi(0)|^2 \le V(t, \phi, \dot{\phi}) \le \gamma \|\phi\|_W^2, \qquad (2.12)$$

$$\dot{V}(t, \phi, \dot{\phi}) + 2\delta V(t, \phi, \dot{\phi}) \le 0. \qquad (2.13)$$

Then (2.5) is exponentially stable with the decay rate δ and (2.11) holds with $K = \frac{\gamma}{\beta}$.

Proof. As in the case of ODEs, from (2.13) with $\phi = x^t$, one derives

$$\frac{d}{dt}V(t, x^t, \dot{x}^t) + 2\delta V(t, x^t, \dot{x}^t) \le 0,$$

where $V(t_0, x^{t_0}, \dot{x}^{t_0}) = V(t_0, \phi, \dot{\phi})$. Hence, by the comparison principle argument [71], it follows that

$$\beta|x(t)|^2 \le V(t, x^t, \dot{x}^t) \le V(t_0, \phi, \dot{\phi})e^{-2\delta(t-t_0)}$$
$$\le \gamma e^{-2\delta(t-t_0)}\|\phi\|_W^2.$$

2.3 Exponential Stability in a Hilbert Space

In this section, the delay is assumed to be either slow-varying subject to

$$\dot{\tau} \le d < 1 \qquad (2.14)$$

or fast-varying with no restrictions on $\dot{\tau}$. Let **A1–A3** be in force.

Delay-dependent conditions are derived by using a simple Lyapunov–Krasovskii functional, inherited from [58]:

$$\begin{aligned} V(t, x^t, \dot{x}^t) &= \langle x(t), Px(t)\rangle + \int_{t-h}^{t} e^{2\delta(s-t)}\langle x(s), Sx(s)\rangle ds \\ &+ h\int_{-h}^{0}\int_{t+\theta}^{t} e^{2\delta(s-t)}\langle \dot{x}(s), R\dot{x}(s)\rangle ds d\theta \\ &+ \int_{t-\tau(t)}^{t} e^{2\delta(s-t)}\langle x(s), Qx(s)\rangle ds, \end{aligned} \qquad (2.15)$$

where $P : \mathcal{D}(A) \to \mathcal{H}$ is a linear positive definite operator and $R, Q, S \in \mathcal{L}(\mathcal{H})$ are nonnegative definite operators, satisfying the following inequalities:

$$\begin{aligned} \beta\langle x, x\rangle &\le \langle x, Px\rangle \le \gamma_P[\langle x, x\rangle + \langle Ax, Ax\rangle], \\ \langle x, Qx\rangle &\le \gamma_Q\langle x, x\rangle, \quad \langle x, Rx\rangle \le \gamma_R\langle x, x\rangle, \\ \langle x, Sx\rangle &\le \gamma_S\langle x, x\rangle, \quad \forall x \in D(A), \end{aligned} \qquad (2.16)$$

for some positive constants $\beta, \gamma_P, \gamma_Q, \gamma_S, \gamma_R$. Thus, condition (2.12) of Lemma 4 is satisfied.

Note that the first inequality (2.16) allows one to use not only bounded operators P (like those considered in [35, p. 217]), but also unbounded operators P, which are upper-estimated by the unbounded operator A according to (2.16). In the case of ODEs where A is a matrix, the above upper bound is equivalent to the standard one with $A = 0$. For ODEs with delay, the Lyapunov functional of the form (2.15) was recently introduced in [60] for $\delta = 0$, whereas this functional with $S = 0$ was introduced earlier in [52] for $\delta = 0$ and in [118] for $\delta > 0$.

Viewed on solutions of (2.5), the Lyapunov–Krasovskii functional (2.15) is absolutely continuous as a function of t because the solutions are absolutely continuous in t. Differentiating V yields

$$\begin{aligned}
&\dot{V}(t, x^t, \dot{x}^t) + 2\delta V(t, x^t, \dot{x}^t) \\
&\leq 2\langle x(t), P\dot{x}(t)\rangle + 2\delta\langle x(t), Px(t)\rangle + h^2 \langle \dot{x}(t), R\dot{x}(t)\rangle \\
&\quad - he^{-2\delta h} \int_{t-h}^{t} \langle \dot{x}(s), R\dot{x}(s)\rangle ds + \langle x(t), (Q+S)x(t)\rangle \\
&\quad - (1 - \dot{\tau}(t))\langle x(t - \tau(t)), Qx(t - \tau(t))\rangle e^{-2\delta h} \\
&\quad - \langle x(t-h), Sx(t-h)\rangle e^{-2\delta h}.
\end{aligned} \quad (2.17)$$

Following [60], the Jensen inequality (2.4), applied to the representation

$$\begin{aligned}
-h \int_{t-h}^{t} \langle \dot{x}(s), R\dot{x}(s)\rangle ds &= -h \int_{t-h}^{t-\tau(t)} \langle \dot{x}(s), R\dot{x}(s)\rangle ds \\
&\quad -h \int_{t-\tau(t)}^{t} \langle \dot{x}(s), R\dot{x}(s)\rangle ds,
\end{aligned} \quad (2.18)$$

results in

$$\begin{aligned}
&\int_{t-\tau(t)}^{t} \langle \dot{x}(s), R\dot{x}(s)\rangle ds \\
&\geq \frac{1}{h} \left\langle \int_{t-\tau(t)}^{t} \dot{x}(s)ds, R \int_{t-\tau(t)}^{t} \dot{x}(s)ds \right\rangle, \\
&\int_{t-h}^{t-\tau(t)} \langle \dot{x}(s), R\dot{x}(s)\rangle ds \\
&\geq \frac{1}{h} \left\langle \int_{t-h}^{t-\tau(t)} \dot{x}(s)ds, R \int_{t-h}^{t-\tau(t)} \dot{x}(s)ds \right\rangle.
\end{aligned} \quad (2.19)$$

Then, taking into account (2.14) and following [55] lead to

$$\begin{aligned}
&\dot{V}(t, x^t, \dot{x}^t) + 2\delta V(t, x^t, \dot{x}^t) \\
&\leq 2\langle x(t), P\dot{x}(t)\rangle + 2\delta\langle x(t), Px(t)\rangle + h^2\langle \dot{x}(t), R\dot{x}(t)\rangle \\
&\quad - \big[\langle x(t) - x(t - \tau(t)), R(x(t) - x(t - \tau(t)))\rangle \\
&\quad + \langle x(t - \tau(t)) - x(t-h), R(x(t - \tau(t)) - x(t-h))\rangle \\
&\quad + (1-d)\langle x(t - \tau(t)), Qx(t - \tau(t))\rangle \big] e^{-2\delta h} \\
&\quad + \langle x(t), (Q+S)x(t)\rangle - \langle x(t-h), Sx(t-h)\rangle e^{-2\delta h}.
\end{aligned} \quad (2.20)$$

Stability conditions are now derived in two forms. The first form will subsequently be applied to the wave equation and the second one to the heat equation. The first form is derived by substituting the right-hand side of (2.5) for $\dot{x}(t)$. Once condition (2.13) of Lemma 4 is given in terms of $\eta(t) = col\{x(t), x(t-h), x(t-\tau(t))\}$, thus yielding

2.3 Exponential Stability in a Hilbert Space

$$\dot{V}(t, x^t, \dot{x}^t) + 2\delta V(t, x^t, \dot{x}^t) \le \langle \eta(t), \Phi_h \eta(t) \rangle \le 0, \quad (2.21)$$

it is readily verified if the following LOI:

$$\Phi_h = \begin{bmatrix} \Phi_{11} & 0 & PA_1 \\ 0 & 0 & 0 \\ A_1^* P & 0 & 0 \end{bmatrix} + h^2 \begin{bmatrix} A^*RA & 0 & A^*RA_1 \\ 0 & 0 & 0 \\ A_1^*RA & 0 & A_1^*RA_1 \end{bmatrix}$$

$$-e^{-2\delta h} \begin{bmatrix} R & 0 & -R \\ 0 & (S+R) & -R \\ -R & -R & 2R+(1-d)Q \end{bmatrix} \le 0 \quad (2.22)$$

holds provided that

$$\Phi_{11} = A^*P + PA + 2\delta P + Q + S. \quad (2.23)$$

The resulting inequality (2.22) is convex with respect to h; that is, given $h_0 > 0$, it becomes feasible for all $\bar{h} \in [0, h_0]$ whenever it is feasible for h_0. The convexity follows from the fact that $\Phi_{\bar{h}} \le \Phi_{h_0}$ since h^2 and $-e^{-2\delta h}$ multiply the nonnegative definite operators. Summarizing, the following result is obtained.

Theorem 5. *Let A1–A3 be in force. Given $\delta > 0$, let there exist linear operators $P > 0$ and $R \ge 0, S \ge 0, Q \ge 0$ subject to (2.16) such that the LOI (2.22) with notation (2.23) holds in the Hilbert space $\mathcal{D}(A) \times \mathcal{D}(A) \times \mathcal{D}(A)$. Then system (2.5) is exponentially stable with the decay rate δ for all slow-varying differentiable delays (2.14). The inequality (2.11) is satisfied with $K = \max\{\gamma_P, h(\gamma_Q + \gamma_S + h^2\gamma_R/2)\}/\beta$. Moreover, (2.5) is exponentially stable for all fast-varying delays $\tau(t)$ with no restrictions on $\dot{\tau}$ if the LOI (2.22) is feasible with $Q = 0$.*

The conditions of Theorem 5 are delay-dependent (h-dependent) even for $\delta \to 0$. Taking in the above derivations $S = R = 0$, one arrives at the following "quasi-delay-independent" conditions, which become delay-independent for $\delta \to 0$; in the case of ODEs, they are simplified to the well-known result.

Corollary 1 ([83]). *Let A1–A3 be in force. Given $\delta > 0$, system (2.5) is exponentially stable with the decay rate δ for all differentiable slow-varying delays (2.14) if there exist linear operators $P > 0$ and $Q \ge 0$ subject to (2.16) such that the LOI*

$$\begin{bmatrix} (A+\delta)^*P + P(A+\delta) + Q & PA_1 \\ A_1^*P & -(1-d)Qe^{-2\delta h} \end{bmatrix} \le 0 \quad (2.24)$$

holds in the Hilbert space $\mathcal{D}(A) \times \mathcal{D}(A)$. Moreover, the inequality (2.11) is satisfied with $K = \max\{\gamma_P, h\gamma_Q\}/\beta$.

Unlike the finite-dimensional case, the feasibility of the strict LOI (2.22) [(2.24)] for $h = 0$ ($\delta = 0$) does not necessarily imply the feasibility of (2.22) [(2.24)] for small enough h (δ) because h^2 (δ) is multiplied by the operator, which may be unbounded.

It may be difficult to verify the feasibility of (2.22) if the operator that multiplies h^2 (and depends on A) in Φ_h is unbounded. To avoid this, we will derive the second form of LOI by the descriptor method [48], where the right-hand sides of the expressions

$$0 = 2\langle x(t), P_2^*[Ax(t) + A_1 x(t - \tau(t)) - \dot{x}(t)]\rangle, \quad (2.25)$$
$$0 = 2\langle \dot{x}(t), P_3^*[Ax(t) + A_1 x(t - \tau(t)) - \dot{x}(t)]\rangle$$

with some $P_2, P_3 \in \mathcal{L}(\mathcal{H})$ are added into the right-hand side of (2.20). Setting $\eta_d(t) = \mathrm{col}\{x(t), \dot{x}(t), x(t-h), x(t-\tau(t))\}$, we obtain

$$\dot{V}(t, x^t, \dot{x}^t) + 2\langle x(t), P\dot{x}(t)\rangle \le \langle \eta_d(t), \Phi_d \eta_d(t)\rangle \le 0,$$

if the LOI

$$\Phi_d = \begin{bmatrix} \Phi_{d11} & \Phi_{d12} & 0 & P_2^* A_1 + Re^{-2\delta h} \\ * & \Phi_{d22} & 0 & P_3^* A_1 \\ * & * & -(S+R)e^{-2\delta h} & Re^{-2\delta h} \\ * & * & * & -[2R + (1-d)Q]e^{-2\delta h} \end{bmatrix} \le 0 \quad (2.26)$$

holds, where

$$\Phi_{d11} = A^* P_2 + P_2^* A + 2\delta P + Q + S - Re^{-2\delta h}, \quad (2.27)$$
$$\Phi_{d12} = P - P_2^* + A^* P_3, \quad \Phi_{d22} = -P_3 - P_3^* + h^2 R$$

and $*$ denotes the symmetric terms of the operator matrix. Thus, the following result is obtained.

Theorem 6. *Let A1–A3 be in force. Given $\delta > 0$, let there exist $P > 0$ and $R \ge 0, S \ge 0, Q \ge 0$ subject to (2.16) and indefinite operators $P_2, P_3 \in \mathcal{L}(\mathcal{H})$ such that the LOI (2.26) with notations given in (2.27) holds in the Hilbert space $\mathcal{D}(A) \times \mathcal{D}(A) \times \mathcal{D}(A) \times \mathcal{D}(A)$. Then system (2.5) is exponentially stable with the decay rate δ for all differentiable slow-varying delays (2.14). The inequality (2.11) is satisfied with $K = \max\{\gamma_P, h(\gamma_Q + \gamma_S + h^2 \gamma_R/2)\}/\beta$. Moreover, (2.5) is exponentially stable for all fast-varying delays with no restrictions on the delay derivative if the LOI (2.26) is feasible with $Q = 0$.*

Consider now system (2.5) with A and A_1 from the uncertain time-invariant polytope

$$\Omega = \sum_{j=1}^M f_j \Omega_j \text{ for some } 0 \le f_j \le 1, \ \sum_{j=1}^M f_j = 1, \quad (2.28)$$

where $\Omega_j = \begin{bmatrix} A^{(j)} & A_1^{(j)} \end{bmatrix}$, $A_1^{(j)} \in \mathcal{L}(\mathcal{H})$ and the operators $A^{(j)}$ have a common domain, which is dense in $\mathcal{H}$ and $A = \sum_{j=1}^{M} f_j A^{(j)}$ generates a strongly continuous semigroup for all f_j, satisfying (2.28). Applying conditions of Theorem 5 to the uncertain system, one concludes that (2.5) is exponentially stable under **A3**, provided LOI (2.22) is feasible. Since LOI (2.22) is affine in A and A_1, by applying the same arguments as for LMIs (see [23]), we may conclude that (2.22) is feasible if the following LOIs:

$$\begin{bmatrix} \Phi_{11} & 0 & PA_1^{(j)} \\ 0 & 0 & 0 \\ A_1^{(j)*}P & 0 & 0 \end{bmatrix} + h^2 \begin{bmatrix} A^{(j)*}RA^{(j)} & 0 & A^{(j)*}RA_1^{(j)} \\ 0 & 0 & 0 \\ A_1^{(j)*}RA^{(j)} & 0 & A_1^{(j)*}RA_1^{(j)} \end{bmatrix}$$

$$-e^{-2\delta h} \begin{bmatrix} R & 0 & -R \\ 0 & (S^{(j)}+R) & -R \\ -R & -R & 2R+(1-d)Q^{(j)} \end{bmatrix} \leq 0,$$

$$\Phi_{11} = A^{(j)*}P + PA^{(j)} + 2\delta P + Q^{(j)} + S^{(j)}, \quad j = 1,\ldots,M,$$

in the vertices are feasible for the same $P > 0$, $R > 0$ and for different $Q^{(j)}$, $S^{(j)}$.

Similarly, if we apply Theorem 6, system (2.5) proves to be exponentially stable provided that LOIs (2.26) in M vertices are feasible for the same P_2, P_3 and for different $Q^{(j)} \geq 0$, $S^{(j)} > 0$, $R^{(j)} > 0$, $P^{(j)} > 0$, $j = 1,\ldots,M$.

2.4 Exponential Stability of a Delay Heat Equation

Consider the heat equation

$$z_t(\xi,t) = az_{\xi\xi}(\xi,t) - a_0 z(\xi,t) - a_1 z(\xi, t - \tau(t)), \tag{2.29}$$

where $t \geq t_0$, $0 \leq \xi \leq l$, with the constant parameters $a > 0$, a_0, and a_1, with the time-varying delay $\tau(t)$, satisfying (2.6), and with the Dirichlet boundary condition

$$z(0,t) = z(l,t) = 0, \ t \geq t_0. \tag{2.30}$$

The boundary-value problem (2.29), (2.30) describes the propagation of heat in a homogeneous one-dimensional rod with a fixed temperature at the ends in the case of the delayed (possibly, due to actuation) heat exchange with the surroundings. Here a and a_i, $i = 0, 1$, stand for the heat conduction coefficient and for the coefficients of the heat exchange with the surroundings, respectively, $z(\xi,t)$ is the value of the temperature field of the plant at time moment t and location ξ along the rod. In the sequel, the state dependence on time t and spatial variable ξ is suppressed whenever possible.

Stability issues of the boundary-value problem (2.29), (2.30) are studied in the Hilbert space $\mathcal{H} = L_2(0, l)$ so that it is represented as the differential equation (2.5) in the Hilbert space $L_2(0, l)$ with the infinitesimal operator $A = a\frac{\partial^2}{\partial \xi^2} - a_0$, possessing the dense domain

$$\mathcal{D}\left(\frac{\partial^2}{\partial \xi^2}\right) = \{z \in W^{2,2}([0, l], \mathbb{R}) : z(0) = z(l) = 0\}, \tag{2.31}$$

and with the bounded operator $A_1 = -a_1$ of the multiplication by the constant $-a_1$. The infinitesimal operator A generates an exponentially stable semigroup (see, e.g., [35] for details).

Simple delay-independent conditions, based on LOI (2.24), are derived first. For this purpose, let's consider the Lyapunov–Krasovskii functional of the form

$$V(t, z^t) = p \int_0^l z^2(\xi, t) d\xi \\ + q \int_{t-\tau(t)}^t \int_0^l e^{2\delta(s-t)} z^2(\xi, s) d\xi ds \tag{2.32}$$

with some positive constants p and q. Then the operators P and Q in (2.24) take the form $P = p$, $Q = q$ of the bounded operators of the multiplication by positive constants p and q, respectively. Integrating by parts and taking into account (2.30) yield

$$\langle x, (A^*P + PA)x \rangle = 2a \int_0^l pzz_{\xi\xi} d\xi - 2a_0 \int_0^l pz^2 d\xi \\ = -2\left[a \int_0^l pz_\xi^2 d\xi + a_0 \int_0^l pz^2 d\xi\right] \leq -2\left(\frac{\pi^2}{l^2}a + a_0\right) \int_0^l pz^2 d\xi \tag{2.33}$$

for $x \in \mathcal{D}(A)$, where the last inequality follows from Wirtinger's inequality (2.1). We thus obtain that (2.24) is satisfied if

$$\Psi_\delta \triangleq \begin{bmatrix} q - 2\left(\frac{\pi^2}{l^2}a + a_0 - \delta\right)p & -a_1 p \\ -a_1 p & -(1-d)qe^{-2\delta h} \end{bmatrix} < 0. \tag{2.34}$$

Since $\Psi_\delta = \Psi_0 + diag\{2\delta p, (1-d)q(1-e^{-2\delta h})\}$, it follows from $\Psi_0 < 0$ that $\Psi_\delta < 0$ for sufficiently small δ. The following result is concluded.

Theorem 7. *Given $\delta > 0$, let the LMI (2.34) hold for some scalars $p > 0$ and $q > 0$. Then the Dirichlet boundary-value problem (2.29), (2.30) is exponentially stable with the decay rate δ for all differentiable slow-varying delays subject to (2.6) and (2.14), and the inequality*

$$\int_0^l z^2(\xi, t) d\xi \leq K e^{-2\delta(t-t_0)} \max_{s \in [t_0-h, t_0]} \int_0^l z^2(\xi, s) d\xi \tag{2.35}$$

is satisfied for all $t \geq t_0$ with $K = 1 + hq/p$. If (2.34) holds for $\delta = 0$, then inequality (2.35) is satisfied with $K = 1 + hq/p$ and a sufficiently small δ.

2.4 Exponential Stability of a Delay Heat Equation

By the Schur complements' formula, LMI (2.34) with $\delta = 0$ is feasible iff $q^2 - 2\left(\frac{\pi^2}{l^2}a + a_0\right)pq + a_1^2 p^2/(1-d) < 0$ for some $p > 0$ and $q > 0$. The left part of the latter inequality achieves its minimum at $q = \left(\frac{\pi^2}{l^2}a + a_0\right)p$ and, thus, the inequality holds iff

$$\frac{\pi^2}{l^2}a + a_0 > 0, \quad a_1^2 < \left(\frac{\pi^2}{l^2}a + a_0\right)^2 (1-d). \tag{2.36}$$

Next, we'll apply Theorem 6 to derive delay-dependent conditions. For simplicity, for this investigation, we've chosen the case where $l = \pi$. The Lyapunov–Krasovskii functional V is thus specified in the form

$$\begin{aligned} V(t, z^t, z_s^t) &= (p_1 - p_3 a) \int_0^\pi z^2(\xi, t) d\xi + p_3 a \int_0^\pi z_\xi^2(\xi, t) d\xi \\ &+ \int_0^\pi \Big[r \int_{-h}^0 \int_{t+\theta}^t e^{2\delta(s-t)} z_s^2(\xi, s) ds d\theta \\ &+ s \int_{t-h}^t e^{2\delta(s-t)} z^2(\xi, s) ds + q \int_{t-\tau(t)}^t e^{2\delta(s-t)} z^2(\xi, s) ds \Big] d\xi, \end{aligned}$$

with some constants $p_1 > 0$, $p_3 > 0$, $s > 0$, $r > 0$, and $q \geq 0$. Then the operators in (2.15) are as follows: $P = -p_3 \left(a \frac{\partial^2}{\partial \xi^2} + a\right) + p_1$, $R = r$, $Q = q$, $S = s$. Furthermore, $P_2 = p_2$ and $P_3 = p_3$ are chosen, where $p_2 > 0$ and

$$p_2 - \delta p_3 \geq 0. \tag{2.37}$$

Here P is an unbounded operator and the other operators are bounded in $L_2(0, \pi)$. It should be noted that the above choice of P, depending on the slack variable P_3, is different from that of the ODEs (where these matrices are independent). Thus, the slack variable allows one to construct an appropriate Lyapunov–Krasovskii functional.

Integrating by parts and utilizing Wirtinger's inequality (2.1), we find that

$$\begin{aligned} \langle x, Px \rangle &= \int_0^\pi \left[-p_3 a z_{\xi\xi} z - p_3 a z^2 + p_1 z^2 \right] d\xi \\ &= \int_0^\pi \left[a p_3 [z_\xi^2 - z^2] + p_1 z^2 \right] d\xi \geq p_1 \int_0^\pi z^2 d\xi > 0 \end{aligned}$$

for $x \in D(A)$. Moreover, (2.16) is satisfied because the inequality

$$\langle x, Px \rangle \leq \int_0^\pi [a p_3 (z_{\xi\xi})^2 + (a p_3 + p_1) z^2] d\xi \leq \gamma_P [|Ax|^2 + |x|^2]$$

holds for some $\gamma_P > 0$ by virtue of the generalized Wirtinger inequality (2.3). We thus obtain that

$$\langle \dot{x}, (P - P_2^* + A^*P_3)x \rangle = \langle \dot{x}, (p_1 - p_2 - (a+a_0)p_3)x \rangle;$$
$$\langle x, A^*P_2 x \rangle + \langle x, P_2^* Ax \rangle + 2\delta \langle x, Px \rangle$$
$$= 2a(p_2 - \delta p_3) \int_0^\pi z_{\xi\xi} z d\xi + 2[-a_0 p_2 + \delta(p_1 - ap_3)] \int_0^\pi z^2 d\xi$$
$$= -2a(p_2 - \delta p_3) \int_0^\pi z_\xi^2 d\xi + 2[-a_0 p_2 + \delta(p_1 - ap_3)] \int_0^\pi z^2 d\xi$$
$$\leq [-2(a+a_0)p_2 + 2\delta p_1] \int_0^\pi z^2 d\xi,$$

where the latter inequality follows from (2.37) and Wirtinger's inequality (2.1). Therefore, (2.26) holds if

$$\begin{bmatrix} \phi_{11} & \phi_{12} & 0 & \phi_{14} \\ * & -2p_3 + h^2 r & 0 & -p_3 a_1 \\ * & * & -(s+r)e^{-2\delta h} & re^{-2\delta h} \\ * & * & * & \phi_{44} \end{bmatrix} < 0, \quad (2.38)$$

$$\phi_{11} = -2(a+a_0)p_2 + 2\delta p_1 + q + s - re^{-2\delta h},$$
$$\phi_{12} = p_1 - p_2 - (a+a_0)p_3,$$
$$\phi_{14} = -p_2 a_1 + re^{-2\delta h}, \quad \phi_{44} = -[2r + (1-d)q]e^{-2\delta h}.$$

We have thus proved the following result.

Theorem 8. *Given $\delta > 0$, let there exist scalars $p_1 > 0, p_2 > 0, p_3 > 0, s > 0, r > 0$, and $q \geq 0$ such that LMIs (2.37) and (2.38) hold. Then the boundary-value problem (2.29), (2.30), where $l = \pi$, is exponentially stable with the decay rate δ for all differentiable slow-varying delays subject to (2.6) and (2.14), and the inequality*

$$p_1 \int_0^\pi z^2(\xi, t) d\xi \leq e^{-2\delta(t-t_0)} \Big\{ ap_3 \int_0^\pi z_\xi^2(\xi, t_0) d\xi$$
$$+ \max[p_1 - p_3 a + hq + hs, h^3 r/2] \qquad (2.39)$$
$$\times \max_{s \in [t_0 - h, t_0]} \int_0^\pi [z^2(\xi, s) + z_t^2(\xi, s)] d\xi \Big\}$$

is satisfied for all $t \geq t_0$. Moreover, (2.29), (2.30) is exponentially stable with the decay rate δ for all fast-varying delays (2.6) with no restrictions on the derivative of the delay if (2.37), (2.38) are feasible with $q = 0$. If (2.38) holds for $\delta = 0$, then (2.29), (2.30) is exponentially stable with a sufficiently small decay rate.

It is of interest to note that under $l = \pi$, the same LMIs (2.34) and (2.38) guarantee the exponential stability of the scalar ODE

$$\dot{y}(t) + (a + a_0) y(t) + a_1 y(t - \tau(t)) = 0. \qquad (2.40)$$

System (2.40) corresponds to the first modal dynamics, corresponding to $k = 1$ in the modal representation

$$\dot{y}_k(t) + (ak^2 + a_0) y_k(t) + a_1 y_k(t - \tau(t)) = 0, \quad k = 1, 2, \ldots, \qquad (2.41)$$

2.4 Exponential Stability of a Delay Heat Equation

of the Dirichlet boundary-value problem (2.29), (2.30) with $l = \pi$, projected on the eigenfunctions of the operator $\frac{\partial^2}{\partial \xi^2}$ (this operator has eigenvalues $-k^2$; see, e.g., [135]). The stability of (2.29), (2.30) implies the stability of (2.41). Thus, the reduction of the infinite-dimensional LOI of Corollary 1 (Theorem 2) to the finite-dimensional LMI of Theorem 7 (Theorem 8) is tight, since the stability of (2.40) is necessary for the stability of (2.29), (2.30).

The above is consistent with the frequency-domain analysis in the case of constant delays, where the characteristic equations of (2.29), (2.30) are given by

$$\lambda_k + ak^2 + a_0 + a_1 e^{-\lambda_k \tau} = 0, \quad k = 1, 2, \ldots \quad (2.42)$$

(see, e.g., [135]). The exponential stability of (2.29), (2.30) is shown in [64] to be determined by (2.42) with $k = 1$, namely, by the stability of the ODE (2.40).

Remark 2. Consider the Dirichlet boundary-value problem (2.29), (2.30) with the uncertain coefficients from the uncertain time-invariant polytope Ω given by (2.28) with $\Omega_j = \begin{bmatrix} a^{(j)} & a_0^{(j)} & a_1^{(j)} \end{bmatrix}$. Here $M = 2^k$ and k is the number of uncertain parameters, and it may take values from the finite set $\{1, 2, 3\}$. The uncertain infinitesimal operator $A = \sum_{j=1}^{M} f_j a^{(j)} \frac{\partial^2}{\partial \xi^2} - a_0^{(j)}$ with the dense domain (2.31) generates a strongly continuous semigroup, whereas the uncertain operator $A_1 = \sum_{j=1}^{M} f_j a^{(j)}$ is bounded. By applying Theorem 8, one concludes that similar to the Hilbert space-valued dynamics, the boundary problem (2.29), (2.30) is exponentially stable if (2.37) holds and LMIs (2.38) in the vertices are feasible for the same p_2, p_3 and for different $q^{(j)} \geq 0, s^{(j)} > 0, r^{(j)} > 0, p^{(j)} > 0, j = 1, \ldots, M$. By Theorem 7, (2.29), (2.30) is exponentially stable if LMIs (2.34) in the vertices are feasible for the same $p > 0$ and for different $q^{(j)}, j = 1, \ldots, M$.

Example 3. To exemplify the above theoretical results, consider the controlled heat equation

$$z_t(\xi, t) = z_{\xi\xi}(\xi, t) + rz(\xi, t) + u, \quad z(0, t) = z(l, t) = 0, \quad (2.43)$$

where $\xi \in (0, l)$, $t > 0$, and where r is an uncertain parameter satisfying $|r| \leq \beta$ with given β. It was shown in [102] that for $l = 1$, the state feedback $u = -\gamma z(\xi, t)$ with $\gamma > \left(\frac{\beta}{2\pi}\right)^2$ exponentially stabilizes (2.43). By verifying the LMI (2.34), we conclude that the closed-loop system is exponentially stable if there exists $p > 0$ such that $-2(\pi^2 - r + \gamma)p < 0$ for all $|r| \leq \beta$, that is, if $\gamma > \beta - \pi^2$. Since $\beta - \pi^2 \leq \left(\frac{\beta}{2\pi}\right)^2$, the developed method guarantees the exponential stabilization of (2.43) via a lower gain, which becomes essentially lower for large β.

Noticing that a time delay often appears in the feedback, let's also consider the case where $l = \pi$, $\beta = 0.1$, and the delayed feedback $u = -z(\xi, t - \tau(t))$ is applied with an uncertain delay satisfying **A3**. This is a polytopic system reached by choosing $r = \pm 0.1$. Applying Theorem 8 with $\delta = 0$ and Remark 2 to the resulting closed-loop system establishes the feasibility of LMI (2.38) in the two vertices corresponding to $r = \pm 0.1$. Given a particular upper bound d in (2.14), we can compute the maximum delay value h_{max} of h in (2.6), for which the closed-loop system remains exponentially stable, by using the MATLAB® LMI toolbox; for example, the toolbox yields $h_{max} = 2.04$ for $d = 0.5$ and $h_{max} = 1.34$ for unknown d. As noted before, these results are inherited from the exponential stability of the ODE $\dot{y} = (-1+r)y(t) - y(t - \tau(t))$ with $|r| \leq 0.1$.

2.5 Exponential Stability of a Delay Wave Equation

Consider the wave equation

$$\begin{aligned} z_{tt}(\xi,t) &= az_{\xi\xi} - \mu_0 z_t(\xi,t) - \mu_1 z_t(\xi, t - \tau(t)) \\ &\quad - a_0 z(\xi,t) - a_1 z(\xi, t - \tau(t)), \quad t \geq t_0, \; 0 \leq \xi \leq \pi \end{aligned} \quad (2.44)$$

with the Dirichlet boundary condition (2.30), where $l = \pi$, and with the constant parameters $a > 0$, $\mu_0 > 0$, μ_1, a_0, and a_1, with the time-varying delay $\tau(t)$, satisfying (2.6). The boundary-value problem (2.30), (2.44) describes the oscillations of a homogeneous string with fixed ends in the case of the delayed (possibly, due to actuation) stiffness restoration and dissipation. Here a stands for the elasticity coefficient, μ_0 and μ_1 stand for the dissipation coefficients, a_0, a_1 stand for the restoring stiffness coefficients, and the state vector $x = col\{z, z_t\}$ consists of the deflection $z(\xi, t)$ of the string and its velocity $z_t(\xi, t)$ at time moment t and location ξ along the string.

Let's introduce the operators

$$A = \begin{bmatrix} 0 & 1 \\ a\frac{\partial^2}{\partial \xi^2} - a_0 & -\mu_0 \end{bmatrix}, \quad A_1 = \begin{bmatrix} 0 & 0 \\ -a_1 & -\mu_1 \end{bmatrix}, \quad (2.45)$$

where A_1 is a bounded operator of multiplication by the constant matrix and where the domain $\mathcal{D}\left(\frac{\partial^2}{\partial \xi^2}\right)$ of the double-differentiation operator is determined by (2.31). Then the boundary-value problem (2.30), (2.44) can be represented as the differential equation (2.5) in the Hilbert space $\mathcal{H} = L_2(0, \pi) \times L_2(0, \pi)$ with the infinitesimal operator A, possessing the domain $\mathcal{D}(A) = \mathcal{D}\left(\frac{\partial^2}{\partial \xi^2}\right) \times L_2(0, \pi)$ and generating an exponentially stable semigroup (see, e.g., [35] for details).

Stability issues of the boundary-value problem (2.30), (2.44) are further studied in the Hilbert space $\mathcal{H} = L_2(0, \pi) \times L_2(0, \pi)$. First, quasi-delay-independent conditions are derived by choosing V in the form

2.5 Exponential Stability of a Delay Wave Equation

$$V(t, v^t) = ap_3 \int_0^\pi z_\xi^2(\xi,t)d\xi + \int_0^\pi v^T(\xi,t)P_0 v(\xi,t)d\xi$$
$$+ \int_{t-\tau(t)}^t \int_0^\pi v^T(\xi,s)e^{2\delta(s-t)}Qv(\xi,s)d\xi ds, \quad (2.46)$$
$$P_0 = \begin{bmatrix} p_1 & p_2 \\ p_2 & p_3 \end{bmatrix}, \quad P_w \triangleq P_0 + diag\{ap_3, 0\} > 0, \quad Q \geq 0,$$

where $v^T(\xi,t) = [z(\xi,t) \ z_t(\xi,t)]$. Then the operators P (unbounded) and Q (bounded) in (2.24) are given by

$$P = diag\left\{-ap_3\tfrac{\partial^2}{\partial \xi^2}, 0\right\} + P_0, \quad Q \geq 0. \quad (2.47)$$

Now, integrating by parts and taking into account the inequality $p_3 > 0$ (extracted from $P_w > 0$) and Wirtinger's inequality (2.1) yield

$$\langle x, Px \rangle = \int_0^\pi \left[-ap_3 z_{\xi\xi}z + z^T P_0 z\right]d\xi = a \int_0^\pi p_3(z_\xi)^2 d\xi$$
$$+ \langle x, P_0 x \rangle \geq \langle x, P_w x \rangle \geq \lambda_{min}(P_w)|x|^2 > 0 \quad (2.48)$$

for all $x \in \mathcal{D}(A) \times L_2(0,\pi)$. Moreover, by the generalized Wirtinger inequality (2.3), the following

$$\langle x, Px \rangle \leq \int_0^\pi ap_3(z_{\xi\xi})^2 d\xi + \langle x, P_0 x \rangle \leq \gamma_P(|Ax|^2 + |x|^2) \quad (2.49)$$

holds with some constant $\gamma_P > 0$, and (2.16) is thus satisfied.

Finally, integrating by parts and applying Wirtinger's inequality (2.1) under condition (2.37) result in

$$\langle x, P(A+\delta)x \rangle + \langle x, (A^* + \delta)Px \rangle = \int_0^\pi [z \ z_t]$$
$$\times \left\{ \begin{bmatrix} p_1 - ap_3\tfrac{\partial^2}{\partial \xi^2} & p_2 \\ p_2 & p_3 \end{bmatrix} \begin{bmatrix} \delta & 1 \\ a\tfrac{\partial^2}{\partial \xi^2} - a_0 & -\mu_0 + \delta \end{bmatrix} \right.$$
$$+ \left. \begin{bmatrix} \delta \ a\tfrac{\partial^2}{\partial \xi^2} - a_0 \\ 1 & -\mu_0 + \delta \end{bmatrix} \begin{bmatrix} p_1 - ap_3\tfrac{\partial^2}{\partial \xi^2} & p_2 \\ p_2 & p_3 \end{bmatrix} \right\} \begin{bmatrix} z \\ z_t \end{bmatrix} d\xi$$
$$= -2a(p_2 - p_3\delta) \int_0^\pi (z_\xi)^2 d\xi - \int_0^\pi [z \ z_t]$$
$$\times \begin{bmatrix} -2p_2 a_0 + 2p_1\delta & p_1-(\mu_0 - 2\delta)p_2 - p_3 a_0 \\ p_1 - (\mu_0 - 2\delta)p_2 - p_3 a_0 & 2p_2 - 2(\mu_0 - \delta)p_3 \end{bmatrix}$$
$$\times \begin{bmatrix} z \\ z_t \end{bmatrix} d\xi \leq \int_0^\pi [z \ z_t](P_w C_\delta + C_\delta^T P_w) \begin{bmatrix} z \\ z_t \end{bmatrix} d\xi, \quad (2.50)$$

where

$$C_\delta = \begin{bmatrix} \delta & 1 \\ -a - a_0 & -\mu_0 + \delta \end{bmatrix}. \tag{2.51}$$

Therefore, (2.24) is feasible if the following LMI

$$\Omega_\delta \triangleq \begin{bmatrix} C_\delta^T P_w + P_w C_\delta + Q & P_w A_1 \\ A_1^T P_w & -(1-d)e^{-2\delta h} Q \end{bmatrix} \leq 0 \tag{2.52}$$

is feasible. Since the LMI (2.52) ensures that the (1,1)-term of its left-hand side satisfies

$$-2p_2(a_0 + a) + 2(p_1 + ap_3)\delta + q_1 \leq 0,$$

whereas $q_1 + 2\delta(p_1 + ap_3) \geq 0$, it follows that $p_2 \geq 0$. Hence, (2.37) does not follow from (2.52).

However, it follows from $\Omega_0 < 0$ that $p_2 > 0$, and (2.37) is thus validated for sufficiently small δ. Moreover, since $\Omega_\delta = \Omega_0 + diag\{2\delta P_w, (1-d)Q(1-e^{-2\delta h})\}$, the inequality $\Omega_0 < 0$ ensures that $\Omega_\delta < 0$ for sufficiently small δ. We have thus proved the following result.

Theorem 9. *Given $\delta > 0$, let the LMIs (2.37) and (2.52) hold for some symmetric 2×2-matrices $P_w > 0$ and $Q \geq 0$, where p_2 and p_3 are, respectively, the (1,2)- and (2,2)-terms of P_w. Then the Dirichlet boundary-value problem (2.30), (2.44), specified with $l = \pi$, is exponentially stable with the decay rate δ for all differentiable slow-varying delays subject to (2.6), (2.14), and the inequality*

$$\lambda_{min}(P_w) \int_0^\pi [z^2(\xi,t) + z_t^2(\xi,t)] d\xi \leq e^{-2\delta(t-t_0)} \\ \times \{K_w \max_{s \in [t_0-h,t_0]} \int_0^\pi [z^2(\xi,s) + z_t^2(\xi,s)] d\xi + ap_3 \int_0^\pi z_\xi^2(\xi,t_0) d\xi\} \tag{2.53}$$

is satisfied, with

$$K_w = \lambda_{max}(P_w - diag\{ap_3, 0\}) + h\lambda_{max}(Q) \tag{2.54}$$

for all $t \geq t_0$. If $\Omega_0 < 0$ holds for $\delta = 0$, then (2.30), (2.44) is exponentially stable with a sufficiently small decay rate.

In a particular case, the above theorem admits further specification.

Corollary 2. *Once specified with $l = \pi$, $a = 1$, and $a_0 = a_1 = 0$, the Dirichlet boundary-value problem (2.30), (2.44) is exponentially stable for all bounded differentiable slow-varying delays subject to (2.6), (2.14) if*

$$\mu_1^2 < (1-d)\mu_0^2. \tag{2.55}$$

2.5 Exponential Stability of a Delay Wave Equation

Proof. Since the delay appears only in z_t, the decision variables of the LMI (2.52) can be chosen in the form $P_w = \begin{bmatrix} 1 & 2\delta \\ 2\delta & 1 \end{bmatrix}$, $Q = \begin{bmatrix} 0 & 0 \\ 0 & \mu_0 \end{bmatrix}$, and thus $P_w A_1 = \begin{bmatrix} 0 & -2\mu_1 \delta \\ 0 & -\mu_1 \end{bmatrix}$. Deleting from Ω_δ the column and the row consisting of zero elements yields the matrix

$$\begin{bmatrix} -2\delta & 4\delta^2 - 2\mu_0\delta & -2\mu_1\delta \\ * & -\mu_0 + 6\delta & -\mu_1 \\ * & * & -(1-d)e^{-2\delta h}\mu_0 \end{bmatrix}.$$

Applying the Schur complements' formula to the last column and the last row of this matrix, we may conclude that (2.52) is feasible if the following LMI:

$$\begin{bmatrix} -2\delta + O(\delta^2) & O(\delta) \\ O(\delta) & -\mu_0 + \frac{\mu_1^2}{(1-d)\mu_0} e^{2\delta h} + 6\delta \end{bmatrix} < 0 \qquad (2.56)$$

holds with $|O(\delta^k)| \leq c\delta^k$ ($k = 1, 2$) for some constant $c > 0$ and for all sufficiently small δ. Subject to (2.55), the (2,2)-term of the left-hand side of (2.56) is negative for sufficiently small δ. Then, applying the Schur complements' formula to the second column and the second row of the matrix in (2.56), we obtain the expression of the form $-2\delta + O(\delta^2)$, which is negative for small δ. Therefore, (2.56) and, hence, (2.52) are feasible for small δ. The proof is completed.

In the constant-delay case $d = 0$, the condition $0 \leq \mu_1 < \mu_0$, which ensures the exponential stability of the wave equation with a mixed Dirichlet–Neumann boundary condition and with $a = 1$, $a_0 = a_1 = 0$, was carried out in [85], where it was also shown that if $\mu_1 \geq \mu_0$, there exists a sequence of arbitrary small delays that destabilize the system.

In order to derive delay-dependent stability conditions for (2.44), (2.30) with $\mu_1 = 0$, we utilize the conditions of Theorem 5. Since the delay appears only in z, V is chosen as follows:

$$V = ap_3 \int_0^\pi z_\xi^2(\xi, t) d\xi + \int_0^\pi [z(\xi, t) \ z_t(\xi, t)] P_0$$
$$\times \begin{bmatrix} z(\xi, t) \\ z_t(\xi, t) \end{bmatrix} d\xi + \int_0^\pi \left[hr \int_{-h}^0 \int_{t+\theta}^t z_t^2(\xi, s) e^{2\delta(s-t)} ds d\theta \right.$$
$$\left. + s \int_{t-h}^t z^2(\xi, s) e^{2\delta(s-t)} ds + q \int_{t-\tau}^t z^2(\xi, s) e^{2\delta(s-t)} ds \right] d\xi,$$

$$P_0 = \begin{bmatrix} p_1 & p_2 \\ p_2 & p_3 \end{bmatrix}, \quad P_w = \begin{bmatrix} ap_3 + p_1 & p_2 \\ p_2 & p_3 \end{bmatrix} > 0,$$

where $r > 0$, $s > 0$, $q \geq 0$. Then the operators P, Q, R in (2.22) are given by

$$P = diag\left\{-ap_3 \frac{\partial^2}{\partial \xi^2}, 0\right\} + P_0 > 0, \quad Q = diag\{r, 0\} \geq 0,$$
$$R = diag\{r, 0\} \geq 0, \quad S = diag\{s, 0\} \geq 0.$$

Since $h^2 A^* RA = diag\{0, h^2 r\}$, $h^2 A^* RA_1 = 0$, $h^2 A_1^* RA_1 = 0$, it follows from (2.50) that (2.22) is feasible if the (1,2) and (2,2) terms p_{02}, p_{03} of P_w meet the condition

$$p_{02} - \delta p_{03} \geq 0 \tag{2.57}$$

and the following LMI:

$$\begin{bmatrix} \phi_w & 0 & P_w \begin{bmatrix} 0 \\ -a_1 \end{bmatrix} + \begin{bmatrix} re^{-2\delta h} \\ 0 \end{bmatrix} \\ * & -(s+r)e^{-2\delta h} & re^{-2\delta h} \\ * & * & -(2r + (1-d)q)e^{-2\delta h} \end{bmatrix} < 0 \tag{2.58}$$

is satisfied provided that $\phi_w = C_\delta^T P_w + P_w C_\delta + diag\{q + s - re^{-2\delta h}, h^2 r\}$. The following result is thus obtained.

Theorem 10. *Given $\delta > 0$, let there exist a 2×2-matrix $P_w > 0$ and scalars $q \geq 0, r > 0, s > 0$ such that LMIs (2.57) and (2.58) hold, with p_{02} and p_{03} as, respectively, the (1,2)- and (2,2)-terms of P_w. Then the delay wave equation (2.44) with $\mu_1 = 0$ and with the Dirichlet boundary condition (2.30), corresponding to $l = \pi$, is exponentially stable with the decay rate δ for all differentiable slow-varying delays subject to (2.6), (2.14), and the inequality (2.53) is satisfied with*

$$K_w = \lambda_{max}(P_w - diag\{ap_3, 0\}) + \max\{hq + hs, h^3 r/2\} \tag{2.59}$$

for all $t \geq t_0$. Moreover, if the LMIs (2.37) and (2.58) are feasible with $q = 0$, then the boundary-value problem (2.30), (2.44) is exponentially stable with the decay rate δ for all fast-varying delays (2.6) with no restrictions on the delay derivative. If the LMI (2.58) holds for $\delta = 0$, then the boundary-value problem (2.30), (2.44) is exponentially stable with a sufficiently small decay rate.

As in the case of the delay heat equation, the LMIs (2.52) and (2.58) ensure the exponential stability of the delay ODE $\dot{\bar{z}}(t) = C_0 \bar{z}(t) + A_1 \bar{z}(t - \tau(t))$, $\bar{z}(t) \in \mathbb{R}^2$ or, equivalently, the scalar delay ODE

$$\begin{aligned} \ddot{y}(t) + \mu_0 \dot{y}(t) + \mu_1 \dot{y}(t - \tau(t)) \\ + (a + a_0) y(t) + a_1 y(t - \tau(t)) = 0. \end{aligned} \tag{2.60}$$

2.5 Exponential Stability of a Delay Wave Equation

Provided that $l = \pi$, the ODE (2.60) governs the first modal dynamics of the modal representation

$$\ddot{y}_k(t) + \mu_0 \dot{y}_k(t) + \mu_1 \dot{y}_k(t - \tau(t)) \\ +(ak^2 + a_0)y_k(t) + a_1 y_k(t - \tau(t)) = 0, \ k = 1, 2, \ldots, \quad (2.61)$$

of the Dirichlet boundary-value problem (2.30), (2.44) on the eigenfunctions of the operator $\frac{\partial^2}{\partial \xi^2}$. Hence, the results of Theorems 9 and 10 are tight in the sense that the stability of ODE (2.60) is necessary for the stability of (2.30), (2.44).

One can also derive "mixed" stability conditions for the wave equation (2.44) with $\mu_1 \neq 0$: delay-dependent (with respect to delay in z)/delay-independent (with respect to delay in z_t). This is similar to stability analysis of neutral systems, where the delay in the state derivative is treated in the delay-independent manner [86].

Similar to Remark 2, the exponential stability of the uncertain wave equation with the coefficients from the uncertain polytope and with the Dirichlet boundary conditions can be verified by solving the LMIs of Theorems 9 and 10 in the vertices of the polytope.

Example 4. To this end, consider the controlled wave equation

$$z_{tt}(\xi, t) = 0.1 z_{\xi\xi}(\xi, t) - 2 z_t(\xi, t) + u, \quad (2.62)$$

with the boundary condition (2.30) and $l = \pi$. By applying Theorem 10, we establish the open-loop system (2.62) with $u = 0$ to be exponentially stable with the decay rate $\delta = 0.05$. Involving a delayed feedback $u = -z(\xi, t - \tau(t))$ and then verifying the conditions of Theorem 10 show that the closed-loop system is exponentially stable with a greater decay rate $\delta = 0.8$ for all $0 \leq \tau(t) \leq 0.31$.

Chapter 3
Linear $\mathcal{H}_\infty$ Control of Time-Varying Systems

The generalization of the $\mathcal{H}_\infty$ synthesis to linear time-varying systems and to linear periodic systems, in particular, is presented here following [100] and [32, 33], respectively.

3.1 Preliminaries

To begin, let's recall basic stability concepts in the time-varying setting. For this purpose, let's consider a linear time-varying system $G(t)$ with the state-space representation

$$\begin{aligned} \dot{x} &= A(t)x + B(t)u, \\ z &= C(t)x + D(t)u, \end{aligned} \qquad (3.1)$$

where $x(t) \in \mathbb{R}^n$, $u(t) \in \mathbb{R}^m$, $z(t) \in \mathbb{R}^l$, and $A(t), B(t), C(t), D(t)$ are piecewise-continuous uniformly bounded matrix functions of appropriate dimensions.

Definition 2. System (3.1) is said to be internally exponentially stable iff the state-transition matrix $\Phi_A(t, \tau)$ of the homogeneous part $\dot{x} = A(t)x$ is such that $\|\Phi_A(t, \tau)\| \leq c_1 e^{-c_2(t-\tau)}$ for all $t \geq \tau$ and some constants $c_1, c_2 > 0$.

As a matter of fact, the above definition is equivalent to writing that any solution $x(t) = \Phi_A(t, \tau)x_0$ of the homogeneous system $\dot{x} = A(t)x$, initialized at $x(\tau) = x_0$, exponentially decays to zero as $t \to \infty$, according to $\|x(t)\| \leq c_1 \|x_0\| e^{-c_2(t-\tau)}$.

Definition 3. System (3.1) is said to be *stabilizable* if there exists a bounded matrix function $K(t)$ such that the system $\dot{x} = (A - BK)(t)x(t)$ is exponentially stable. The notation $(A(t), B(t))$ stabilizable is also used to denote this.

Definition 4. System (3.1) is said to be *detectable* if there exists a bounded matrix function $L(t)$ such that the system $\dot{x} = (A - LC)(t)x(t)$ is exponentially stable. The notation $(C(t), A(t))$ detectable is also used to denote this.

It is worth noticing that if the time-varying system (3.1) is stabilizable and detectable, then it is internally exponentially stable provided that its input–output operator

$$T_{zu}(G)u = \int_0^t C(t)\Phi_A(t,\tau)B(\tau)u(\tau)d\tau + D(t)u(t), \quad (3.2)$$

defined on the $\mathcal{L}_2^m(0, \infty)$-space of m-vector square-integrable functions $u(\cdot)$, is bounded.

Conversely, if system (3.1) is internally exponentially stable, then its input–output operator (3.2) is a bounded linear operator

$$T_{zu}(G) : \mathcal{L}_2^m(0, \infty) \to \mathcal{L}_2^l(0, \infty).$$

With this in mind, the induced norm $\|T_{zu}(G)\| = \sup_{\|u\| \neq 0} \frac{\|T_{zu}u\|}{\|u\|}$ of the input–output operator (3.2) represents a natural extension of the $\mathcal{H}_\infty$ norm (1.13) to a linear time-varying system $G(\cdot)$ with the state-space representation (3.1):

$$\|G(\cdot)\|_\infty \triangleq \|T_{zu}(G)\| = \sup_{\|u\| \neq 0} \frac{\|T_{zu}u\|}{\|u\|}. \quad (3.3)$$

3.1.1 Strict Bounded Real Lemma in Time-Varying and Periodic Settings

For later use, the strict bounded real lemma (see, e.g., [25, Lemma 5.65]) that gives the algebraic characterization of the transfer function to meet the bounded real condition (1.30) is extended to the time-varying setting.

The next lemma is for supporting the proof of the ongoing extension of the strict bounded real lemma to time-varying systems.

Lemma 5. *Let these conditions be satisfied:*

- *A time-dependent $n \times n$-matrix $\mathcal{S}(t)$ is piecewise-continuous, uniformly bounded, symmetric, and positive definite.*
- *A time-dependent $n \times n$-matrix $\mathcal{A}(t)$ is piecewise-continuous and uniformly bounded.*
- *The corresponding differential equation*

$$\dot{x} = \mathcal{A}(t)x \quad (3.4)$$

is uniformly asymptotically stable.

3.1 Preliminaries

Then the matrix

$$Q(t) = \int_t^\infty \Phi_A^T(\tau, t) S(\tau) \Phi_A(\tau, t) d\tau, \tag{3.5}$$

specified with the above matrix $S(t)$ and the transition matrix $\Phi_A(\tau, t)$ of (3.4), is uniformly bounded, symmetric, and positive definite. If, in addition, the matrix functions $S(t)$ and $A(t)$ are T-periodic, then $Q(t)$ given by (3.5) is T-periodic as well.

Proof. The proof of this auxiliary result is rather technical; its details may be found in [125, p. 181]. Here we provide only a sketch.

By virtue of the uniform asymptotic stability of (3.4), the exponential estimate

$$\|\Phi_A(\tau, t)\| \le m e^{-\lambda(\tau - t)} \tag{3.6}$$

of the transition matrix holds for some positive λ and m. Taking this into account and utilizing that $S(t)$ is uniformly bounded, one can then conclude the uniform boundedness of (3.5). It is additionally clear that relation (3.5) defines a symmetric matrix whenever $S(t)$ is so.

Furthermore, let $s(\tau, q, t) = \Phi_A(\tau, t)x$ denote the solution of (3.4), initialized at a time instant t with $x(t) = q \in \mathbb{R}^n$ and evaluated at a time instant τ. By employing the positive definiteness of $S(t)$, we see from (3.5) that

$$q^T Q(t) q = \int_t^\infty s^T(\tau, q, t) S(\tau) s(\tau, q, t) d\tau \ge \alpha \int_t^\infty \|s(\tau, q, t)\|^2 d\tau. \tag{3.7}$$

In turn,

$$\|s(\tau, q, t)\| \ge \|q\| \exp\left\{-\int_t^\tau \mu[-A(\theta)] d\theta\right\}$$

$$\ge \|q\| \exp\left\{-\int_t^\tau \|A(\theta)\| d\theta\right\} \ge \|q\| e^{-m_0(\tau - t)}, \tag{3.8}$$

where $m_0 = \sup_t \|A(t)\|$ is an upper bound of the induced matrix norm of $A(t)$ and $\mu[A(t)] = \lim_{\varepsilon \to 0+} \frac{\|I + \varepsilon A(t)\| - 1}{\varepsilon}$ stands for the matrix measure, as it is the directional derivative of the matrix norm function $\|\cdot\|$ evaluated at the identity matrix I in the direction $A(t)$. Now relations (3.7) and (3.8), coupled together, result in the positive definiteness of matrix (3.5):

$$q^T Q(t) q \ge \alpha \int_t^\infty q^T q e^{-2m_0(\tau - t)} d\tau = \frac{\alpha}{2m_0} q^T q. \tag{3.9}$$

To complete the proof, we note that once the matrix functions $S(t)$ and $\mathcal{A}(t)$ are T-periodic, the transition matrix $\Phi_A(\tau,t)$ is such that $\Phi_A(\tau,t) = \Phi_A(\tau+T, t+T)$ for all $\tau, t \in \mathbb{R}$, and therefore, the function $\mathcal{Q}(t)$ in relation (3.5) proves to be T-periodic, too. Indeed,

$$\mathcal{Q}(t+T) = \int_{t+T}^{\infty} \Phi_A^T(\tau, t+T) \mathcal{S}(\tau) \Phi_A(\tau, t+T) d\tau$$

$$= \int_{t}^{\infty} \Phi_A^T(\zeta+T, t+T) \mathcal{S}(\zeta+T) \Phi_A(\zeta+T, t+T) d\zeta$$

$$= \int_{t}^{\infty} \Phi_A^T(\zeta, t) \mathcal{S}(\zeta) \Phi_A(\zeta, t) d\zeta = \mathcal{Q}(t), \qquad (3.10)$$

where the integration variable substitution $\tau = \zeta + T$ has been applied. Lemma 5 is thus proved.

According to [90, 100], the time-varying version of the strict bounded real lemma reads as follows.

Lemma 6 (Time-varying strict bounded real lemma). *Consider a linear time-varying system $G(\cdot)$ with the state-space representation (3.1), where $A(t) \in \mathbb{R}^{n \times n}$, $B(t) \in \mathbb{R}^{r \times n}$, $C(t) \in \mathbb{R}^{n \times p}$ are piecewise-continuous uniformly bounded matrix functions, and $D(t) \equiv 0$. Then, if we are given $\gamma > 0$, the following statements are equivalent:*

(i) *System (3.1) is internally exponentially stable and its $\mathcal{H}_\infty$ norm (3.3) is less than γ.*
(ii) *The differential Riccati equation (DRE)*

$$\dot{P} + P(t)A(t) + A^T(t)P(t) + C^T(t)C(t) + \frac{1}{\gamma^2} P(t) BB^T(t) P(t) = 0$$

$$(3.11)$$

has a uniformly bounded positive semidefinite solution $P(t) \in \mathbb{R}^{n \times n}$ such that system

$$\dot{x} = [A + \gamma^{-2} BB^T P](t) x \qquad (3.12)$$

is exponentially stable.
(iii) *There exists $\varepsilon_0 > 0$ such that the perturbed DRE*

$$\dot{P}_\varepsilon + P_\varepsilon(t)A(t) + A^T(t)P_\varepsilon(t) + C^T(t)C(t) + \frac{1}{\gamma^2} P_\varepsilon(t) BB^T(t) P_\varepsilon(t) + \varepsilon I = 0$$

$$(3.13)$$

with the identity matrix $I \in \mathbb{R}^{n \times n}$ has a unique uniformly bounded positive definite symmetric solution $P_\varepsilon(t)$ for each $\varepsilon \subset (0, \varepsilon_0)$.

3.1 Preliminaries

Proof. The equivalence of (i) and (ii) is well known from [100]. In order to prove the implication $(ii) \to (iii)$, let us represent Eq. (3.13) in the form

$$\Gamma(P_\varepsilon) + \varepsilon I = 0, \qquad (3.14)$$

where

$$\Gamma : P(t) \to \dot{P} + P(t)A(t) + A^T(t)P(t) + C^T(t)C(t) + \frac{1}{\gamma^2} P(t)B(t)B^T(t)P(t) \qquad (3.15)$$

is a mapping from the space

$$\mathcal{B} = \{P(\cdot) \in L_\infty(-\infty, \infty) : P(\cdot) = P^T(\cdot) \text{ and } \dot{P}(\cdot) \in L_\infty(-\infty, \infty)\} \qquad (3.16)$$

of differentiable uniformly bounded symmetric matrix functions with uniformly bounded derivatives to the space

$$\mathcal{B}_1 = \{P_1(\cdot) \in L_\infty(-\infty, \infty) : P_1(\cdot) = P_1^T(\cdot)\} \qquad (3.17)$$

of uniformly bounded symmetric matrix functions. Clearly, Eq. (3.11) is equivalent to $\Gamma(P_{\varepsilon=0}) = 0$.

First, we demonstrate that the tangent map

$$D\Gamma_P : Q(t) \to \dot{Q}(t) + \left[A(t) + \frac{1}{\gamma^2} B(t)B^T(t)P(t)\right]^T Q(t) +$$

$$Q(t)\left[A(t) + \frac{1}{\gamma^2} B(t)B^T(t)P(t)\right] \qquad (3.18)$$

of Γ at $P_{\varepsilon=0}$ is invertible. In other words, we demonstrate that for any $S(t) \in \mathcal{B}_1$, the equation

$$D\Gamma_P(Q) + S = 0 \qquad (3.19)$$

has a unique solution $Q(t) \in \mathcal{B}$. Indeed, it is straightforward to check that such a solution is given by

$$Q_s(t) = \int_t^\infty \Phi^T(\tau, t) S(\tau) \Phi(\tau, t) d\tau, \qquad (3.20)$$

where $\Phi(\tau, t)$ is the transition matrix of (3.12), which, under the lemma conditions, satisfies the inequality

$$\|\Phi(\tau, t)\| \le m e^{-\lambda(\tau-t)} \qquad (3.21)$$

for all $\tau \geq t$ and some positive m and λ. Moreover, since arbitrary solution of (3.19) admits representation

$$Q(t) = \Phi^T(0,t) Q(0) \Phi(0,t) + \int_0^t \Phi^T(\tau,t) S(\tau) \Phi(\tau,t) d\tau, \quad (3.22)$$

the difference $\Phi^T(0,t)[Q_1(0) - Q_2(0)]\Phi(0,t)$ between two uniformly bounded solutions $Q_1(t)$ and $Q_2(t)$ of (3.19) is uniformly bounded iff these solutions have the same initial conditions. Otherwise, there would exist a nonzero vector $q \in \mathbb{R}^n$, satisfying the inequality

$$\|\Phi(0,t)q\| \leq K$$

with some $K > 0$, which by virtue of (3.21) results in the false statement that $0 \neq \|q\| = \|\Phi(t,0)\Phi(0,t)q\| \leq \|\Phi(t,0)\| \|\Phi(0,t)q\| \leq Kme^{-\lambda t} \to 0$ as $t \to \infty$. This contradiction proves the uniqueness of the uniformly bounded solution of (3.19). Due to Lemma 5, solution (3.20) is positive definite whenever $S(t)$ is positive definite. By the implicit function theorem (see, e.g., [78, p. 194]), it follows that (3.14) has a unique bounded positive definite symmetric solution $P_\varepsilon(t)$ for each $\varepsilon > 0$ small enough. The implication $(ii) \to (iii)$ is thus established.

To complete the proof, we will demonstrate that $(iii) \to (i)$. For this purpose, the Lyapunov function $V(xt) = x^T P_\varepsilon(t)x$ is involved, where according to statement (iii) of the lemma, $P_\varepsilon(t)$ is chosen with sufficiently small $\varepsilon > 0$, and the rest of the proof follows the same line of reasoning as that used in the proof of Theorem 4. The details are left to the reader. Lemma 6 is thus proved.

While being confined to the periodic setting, Lemma 6 is readily modified to the following [30, 109].

Lemma 7 (Periodic strict bounded real lemma). *Consider a linear time-periodic system $G(\cdot)$ with the state-space representation (3.1), where $A(t) \in \mathbb{R}^{n \times n}$, $B(t) \in \mathbb{R}^{r \times n}$, $C(t) \in \mathbb{R}^{n \times p}$ are piecewise-continuous T-periodic matrix functions, and $D(t) \equiv 0$. Then given $\gamma > 0$, the following statements are equivalent:*

(i) *System (3.1) is internally exponentially stable and its $\mathcal{H}_\infty$ norm (3.3) is less than γ.*
(ii) *The periodic DRE (3.11) has a T-periodic positive semidefinite solution $P(t) \in \mathbb{R}^{n \times n}$ such that the periodic system (3.12) is exponentially stable.*
(iii) *There exists $\varepsilon_0 > 0$ such that the perturbed DRE (3.13) with the identity matrix $I \in \mathbb{R}^{n \times n}$ has a unique T-periodic, positive definite symmetric solution $P_\varepsilon(t)$ for each $\varepsilon \subset (0, \varepsilon_0)$.*

Proof. To the periodic case, the equivalence of (i) and (ii) is extended in [137]. To subsequently verify that the implications $(ii) \to (iii)$ and $(iii) \to (i)$ are in force in the periodic case as well, we represent Eq. (3.13) in the form (3.14), where (3.15) is a mapping from the space $\mathcal{T} \subset \mathcal{B}$ of differentiable T-periodic symmetric matrix functions with uniformly bounded derivatives to the space $\mathcal{T}_1 \subset \mathcal{B}_1$ of T-periodic symmetric matrix functions. Clearly, Eq. (3.11) is equivalent to $\Gamma(P_{\varepsilon=0}) = 0$; for us to demonstrate $(ii) \to (iii)$, it remains to prove that this equation has a unique T-periodic, positive definite symmetric solution $P_\varepsilon(t)$ for each $\varepsilon \subset (0, \varepsilon_0)$. In the periodic case, the proof of this fact and that of the implication $(iii) \to (i)$ follow the same line of reasoning used in the proof of Lemma 6, and their details are therefore omitted. This completes the proof of Lemma 7.

3.2 Synthesis of Time-Varying Systems

The state-space representation (1.27) is now considered in the time-varying setting; that is, the system $G(t)$ of interest is governed by

$$\begin{aligned} \dot{x} &= A(t)x + B_1(t)w + B_2(t)u \\ z &= C_1(t)x + D_{12}(t)u \\ y &= C_2(t)x + D_{21}(t)w \end{aligned} \quad (3.23)$$

and an admissible controller (1.28) is of the form

$$\begin{aligned} \dot{\xi} &= A_K(t)\xi + B_{1K}(t)y + B_{2K}(t)u, \\ u &= C_K(t)\xi + D_K(t)y. \end{aligned} \quad (3.24)$$

As before, the state $x(t) \in \mathbb{R}^n$, the control input $u(t) \in \mathbb{R}^m$, the unknown disturbance $w(t) \in \mathbb{R}^r$, the to-be-controlled output $z(t) \in \mathbb{R}^l$, the available measurement $y(t) \in \mathbb{R}^p$, the internal controller state $\xi \in \mathbb{R}^k$, whereas the state matrix functions $A(t), B_1(t), B_2(t), C_1(t), C_2(t), D_{12}(t), D_{21}(t)$ and the controller matrix functions $A_K(t), B_{1K}(t), B_{2K}(t), C_K(t), D_K(t)$ are of appropriate dimensions, piecewise-continuous, and uniformly bounded in t.

The *time-varying $\mathcal{H}_\infty$ optimal control problem*, properly extended from that in the time-invariant setting, is to find, among all admissible controllers (3.24), an internally stabilizing controller $K(t)$ that minimizes the induced $\mathcal{H}_\infty$ norm (3.3), specified in terms of the input–output operator $T_{zw}(G, K)$ of the resulting closed-loop system (3.23), (3.24).

Let γ_{min} be the minimum value of $\|T_{zw}(G, K)\|$ over all internally stabilizing controllers (3.24). Then $\mathcal{H}_\infty$ *suboptimal control problem* is to find an admissible internally stabilizing controller $K(t)$ such that

$$\|(G, K)\|_\infty \triangleq \|T_{zw}(G, K)\|_\infty < \gamma \quad (3.25)$$

for an a priori given disturbance attenuation level $\gamma > \gamma_{min}$.

A suboptimality test, stated below, is similar to that of Theorem 2 and is in force under the simplifying assumptions

($\mathcal{A}$1) $(A(t), B_1(t))$ is stabilizable and $(C_1(t), A(t))$ is detectable,
($\mathcal{A}$2) $(A(t), B_2(t))$ is stabilizable and $(C_2(t), A(t))$ is detectable,
($\mathcal{A}$3) $D_{12}^T(t)C_1(t) \equiv 0$ and $D_{12}^T(t)D_{12}(t) \equiv I$,
($\mathcal{A}$4) $B_1(t)D_{21}^T(t) \equiv 0$ and $D_{21}(t)D_{21}^T(t) \equiv I$

inherited from the time-invariant treatment. Necessary and sufficient conditions for the above $\mathcal{H}_\infty$ suboptimal control problem to have a solution with a disturbance attenuation level $\gamma > 0$ are formulated in terms of the existence of appropriate solutions of certain DREs whose algebraic terms copy those of the algebraic Riccati equations (AREs) (1.31), (1.37):

(S1) The equation

$$-\dot{P} = P(t)A(t) + A^T(t)P(t) + C_1^T(t)C_1(t)$$
$$+ P(t)\left[\frac{1}{\gamma^2}B_1B_1^T - B_2B_2^T\right](t)P(t) \qquad (3.26)$$

possesses a uniformly bounded positive semidefinite symmetric solution $P(t)$ such that the system

$$\dot{x} = [A - (B_2B_2^T - \gamma^{-2}B_1B_1^T)P](t)x(t) \qquad (3.27)$$

is exponentially stable.

(S2) As it is specified with

$$A_1(t) = A(t) + \frac{1}{\gamma^2}B_1(t)B_1^T(t)P(t), \qquad (3.28)$$

the equation

$$\dot{Q} = A_1(t)Q(t) + Q(t)A_1^T(t) + B_1(t)B_1^T(t)$$
$$+ Q(t)\left[\frac{1}{\gamma^2}PB_2B_2^T P - C_2^T C_2\right](t)Q(t) \qquad (3.29)$$

possesses a uniformly bounded positive semidefinite symmetric solution $Q(t)$ such that the system

$$\dot{x} = [A_1 - Q(C_2^T C_2 - \gamma^{-2}PB_2B_2^T P)](t)x(t) \qquad (3.30)$$

is exponentially stable.

3.2 Synthesis of Time-Varying Systems

Along with the suboptimality test, the following result, established in [100], yields a suboptimal controller that by iteration on γ approaches an optimal solution of the time-varying $\mathcal{H}_\infty$ control problem.

Theorem 11. *The $\mathcal{H}_\infty$ suboptimal control problem possesses a solution for the plant model (3.23) with Assumptions (A1)–(A4) and a disturbance attenuation level $\gamma > 0$ if and only if Conditions (S1) and (S2), coupled together, are satisfied. Once these conditions hold, the controller*

$$\dot{\xi} = A(t)\xi(t) + \left(\frac{1}{\gamma^2} B_1 B_1^T - B_2 B_2^T\right)(t) P(t)\xi(t)$$
$$+ Q(t) C_2^T(t)[y(t) - C_2(t)\xi(t)], \quad (3.31)$$
$$u = -B_2^T(t) P(t)\xi(t)$$

yields a solution to the problem in question.

An alternative suboptimality test is obtained if the DREs (3.26) and (3.29) are perturbed with the identity matrix $I \in \mathbb{R}^{n \times n}$ and a positive ε:

$$-\dot{P}_\varepsilon = P_\varepsilon(t) A(t) + A^T(t) P_\varepsilon(t) + C_1^T(t) C_1(t)$$
$$+ P_\varepsilon(t) \left[\frac{1}{\gamma^2} B_1 B_1^T - B_2 B_2^T\right](t) P_\varepsilon(t) + \varepsilon I, \quad (3.32)$$

$$\dot{Q}_\varepsilon = A_1(t) Q_\varepsilon(t) + Q_\varepsilon(t) A_1^T(t) + B_1(t) B_1^T(t)$$
$$+ Q_\varepsilon(t) \left[\frac{1}{\gamma^2} P_\varepsilon B_2 B_2^T P_\varepsilon - C_2^T C_2\right](t) Q_\varepsilon(t) + \varepsilon I. \quad (3.33)$$

To reproduce such a test, we modify Conditions (S1) and (S2) to

(S1) There exists a uniformly bounded positive definite symmetric solution $P_\varepsilon(t)$ of the perturbed DRE (3.32) with some $\varepsilon > 0$.
(S2) There exists a uniformly bounded positive definite symmetric solution $Q_\varepsilon(t)$ of the perturbed DRE (3.33), specified with (3.28) and some $\varepsilon > 0$.

Theorem 12. *The $\mathcal{H}_\infty$ suboptimal control problem possesses a solution for the plant model (3.23) with Assumptions (A1)–(A4) and a disturbance attenuation level $\gamma > 0$ if and only if Conditions (S1) and (S2), coupled together, are satisfied. Once these conditions hold, the controller*

$$\dot{\xi} = A(t)\xi(t) + \left(\frac{1}{\gamma^2} B_1 B_1^T - B_2 B_2^T\right)(t) P_\varepsilon(t)\xi(t)$$
$$+ Q_\varepsilon(t) C_2^T(t)[y(t) - C_2(t)\xi(t)], \quad (3.34)$$
$$u = -B_2^T(t) P_\varepsilon(t)\xi(t)$$

yields a solution to the problem in question.

Proof. *Necessity* is demonstrated by invoking a time-varying version of the strict bounded real lemma. Let the $\mathcal{H}_\infty$ suboptimal control problem possess a solution. Then by Theorem 11, Conditions (*S*1) and (*S*2) hold, and by applying Lemma 6, we see that Conditions (*S*1) and (*S*2) follow.

Sufficiency is proved based on the Lyapunov function,

$$U(x,\xi,t) = x^T P_\varepsilon(t)x + \gamma^2(x-\xi)^T Q_\varepsilon^{-1}(t)(x-\xi),$$

where $P_\varepsilon(t)$ and $Q_\varepsilon(t)$ are uniformly bounded positive definite symmetric solutions of the perturbed DREs (3.32) and (3.33), specified with (3.28) and some $\varepsilon > 0$, whereas the existence of the desired functions $P_\varepsilon(t)$ and $Q_\varepsilon(t)$ is guaranteed by Conditions (*S*1) and (*S*2). The detailed proof of the sufficiency is similar to that of Theorem 4 and is left to the reader.

3.3 Synthesis of Periodic Systems

In the case of a periodic system (3.23), where the matrix functions $A(t), B_1(t), B_2(t), C_1(t), C_2(t), D_{12}(t), D_{21}(t)$ are time-periodic of a period $T > 0$, the interest is focused on the design of an admissible time-periodic controller (3.24). Just in case, Conditions (*S*1) and (*S*2) are modified to require appropriate solutions of the DREs (3.26) and (3.29) to be periodic:

(*T*1) The DRE (3.26) possesses a T-periodic, positive semidefinite symmetric solution $P(t)$ such that (3.27) is exponentially stable.

(*T*2) Specified with (3.28), the DRE (3.29) possesses a T-periodic, positive semidefinite symmetric solution $Q(t)$ such that system (3.30) is exponentially stable.

The following result, extracted from [33, 137], is in order.

Theorem 13. *Consider the time-varying plant model (3.23) with the time-periodic matrices $A(t), B_1(t), B_2(t), C_1(t), C_2(t), D_{12}(t), D_{21}(t)$ of a period $T > 0$ and Assumptions ($\mathcal{A}$1)–($\mathcal{A}$4). Then the corresponding $\mathcal{H}_\infty$ suboptimal control problem possesses a T-periodic solution with a disturbance attenuation level $\gamma > 0$ if and only if Conditions (T1) and (T2), coupled together, are satisfied. Once these conditions hold, controller (3.31) yields a T-periodic solution to the problem in question.*

Similar to Theorem 12, an alternative suboptimality test becomes available in the periodic setting as well once Conditions (T1) and (T2) are reformulated in terms of the perturbed DREs:

($\mathcal{T}$1) There exists a T-periodic, positive definite symmetric solution $P_\varepsilon(t)$ of the perturbed DRE (3.32) with some $\varepsilon > 0$.

3.3 Synthesis of Periodic Systems

(T2) There exists a T-periodic, positive definite symmetric solution $Q_\varepsilon(t)$ of the perturbed DRE (3.33) specified with (3.28) and some $\varepsilon > 0$.

Theorem 14. *Consider the time-varying plant model (3.23) with the time-periodic matrices $A(t), B_1(t), B_2(t), C_1(t), C_2(t), D_{12}(t), D_{21}(t)$ of a period $T > 0$ and Assumptions ($\mathcal{A}$1)–($\mathcal{A}$4). Then the corresponding $\mathcal{H}_\infty$ suboptimal control problem possesses a T-periodic solution with a disturbance attenuation level $\gamma > 0$ if and only if Conditions (T1) and (T2), coupled together, are satisfied. Once these conditions hold, controller (3.34) yields a T-periodic solution to the problem in question.*

Proof. **Necessity** is demonstrated based on a periodic version of the strict bounded real lemma. Let the $\mathcal{H}_\infty$ suboptimal control problem possess a solution. Then by Theorem 13, Conditions (T1) and (T2) hold, and by applying Lemma 7, we establish the validity of Conditions (T1) and (T2).

Sufficiency is proved based on the Lyapunov function

$$U(x, \xi, t) = x^T P_\varepsilon(t) x + \gamma^2 (x - \xi)^T Q_\varepsilon^{-1}(t)(x - \xi)$$

with $P_\varepsilon(t)$ and $Q_\varepsilon(t)$ being T-periodic, positive definite symmetric solutions of the perturbed DREs (3.32) and (3.33), specified with (3.28) and some $\varepsilon > 0$. The existence of the desired functions $P_\varepsilon(t)$ and $Q_\varepsilon(t)$ is guaranteed by Conditions (T1) and (T2). The detailed proof of the sufficiency follows the line of reasoning used in the proof of Theorem 4 and is left to the reader.

Chapter 4
Nonlinear $\mathcal{H}_\infty$ Control

In this chapter, the $\mathcal{H}_\infty$ (sub)optimal control problem is reformulated for an autonomous nonlinear system in terms of its $\mathcal{L}_2$-induced norm. If confined to linear systems, this norm is recognized as the time-domain version of the standard $\mathcal{H}_\infty$ norm [see, e.g., relation (3.3) for the $\mathcal{H}_\infty$-norm extension in the time-varying setting]. An additional motivation comes from the time-domain interpretation, where the $\mathcal{H}_\infty$ norm stands for the maximum gain in the steady-state response to sinusoidal inputs. Such an interpretation is presented here for nonlinear systems as well, and the state-space solutions of the problem are further derived in the nonlinear setting, similar to the approach proposed by Isidori and Astolfi [67] and Van der Schaft [122]. The more general approach with detailed coverage of nonlinear $\mathcal{H}_\infty$ control can be found in the book by Helton and James [61].

4.1 The $\mathcal{L}_2$ Gain and $\mathcal{H}_\infty$ Gain of a Nonlinear System

Consider a nonlinear system of the form

$$\dot{x}(t) = f(x(t)) + g(x(t))w(t), \tag{4.1}$$

$$z(t) = h(x(t)), \tag{4.2}$$

where $x \in \mathbb{R}^n$, $t \in \mathbb{R}$, $w \in \mathbb{R}^r$, and $z \in \mathbb{R}^p$, and $f(x), g(x), h(x)$ are matrix functions of appropriate dimensions. In order to ensure that, first, system (4.1), (4.2) is locally well posed for any locally integrable input $w(\cdot)$ and, second, the origin is an equilibrium point of the nominal system with $w \equiv 0$, the matrix functions $f(x), g(x), h(x)$ are assumed to be smooth (i.e., of class C^∞) and such that $f(0) = 0$ and $h(x) = 0$. The well-known $\mathcal{L}_2$-gain concept that takes its origins in [131] is then specified as follows.

Definition 5. Given a real number $\gamma > 0$, system (4.1) is said to have $\mathcal{L}_2$ *gain less than* γ with respect to output (4.2) [or, simply, system (4.1), (4.2) has $\mathcal{L}_2$ gain less than γ] if a system response z, resulting from w for initial state $x(t_0) = 0$, exists on an arbitrary interval $(0, T)$ and satisfies

$$\int_0^T \|z(t)\|^2 \, dt < \gamma^2 \int_0^T \|w(t)\|^2 \, dt \tag{4.3}$$

for all $T \geq 0$ and all piecewise-continuous functions $w(t)$.

Definition 6. System (4.1), (4.2) is said to have $\mathcal{L}_2$ *gain less than* γ, *locally around the origin*, if there exists a neighborhood U of the origin such that inequality (4.3) is satisfied for all $T \geq 0$ and all piecewise-continuous functions $w(t)$ for which the state trajectory of (4.1) starting from the initial point $x(0) = 0$ remains in U for all $t \in (0, T)$.

The above definitions characterize the gain of the nonlinear system in terms of its $\mathcal{L}_2$-induced norm and may be viewed as a special case of the general definition of dissipativity [132, 133], where an arbitrary supply rate $L(z, w)$ is specified with the quadratic one $L(z, w) = \gamma^2 \|w\|^2 - \|z\|^2$.

For deeper insight into evaluating the nonlinear response to persistent (particularly, periodic) disturbances whose $\mathcal{L}_2$ norm escapes to infinity over infinite intervals, we additionally outline a direct approach of [67] to the nonlinear counterpart of the $\mathcal{H}_\infty$ norm. In the approach, the $\mathcal{H}_\infty$ norm is interpreted as the maximum gain in the steady-state response to sinusoidal inputs and is extended to nonlinear systems. The extension is based on the following observation [68].

Let a nonlinear system

$$\dot{x} = f(x, w), \tag{4.4}$$

whose state $x \in \mathbb{R}^n$ and exogenous input $w \in \mathbb{R}^r$, be locally exponentially stable around the origin $x = 0, w = 0$. Such a system is said to be *locally exponentially stable around the origin* $x = 0, w = 0$ iff the eigenvalues of the Jacobian matrix $\partial f(x, w)/\partial w$, computed at $x = 0, w = 0$, are in the open left half-plane. Apart from this, let the exogenous inputs be generated by a harmonic oscillator or, generally speaking, by a dynamical system

$$\dot{w} = s(w), \tag{4.5}$$

where all trajectories are periodic of a fixed period T. Then the composed system (4.4), (4.5) possesses a *center manifold*

$$S = \{(x, w) : x = \pi(w)\} \tag{4.6}$$

with a C^1 mapping $\pi(w)$, locally defined in neighborhood W of the point $w = 0$, such that $\pi(0) = 0$ and $\pi'(0) = 0$, where $\pi'(w) = \partial \pi/\partial w$. The center manifold

4.1 The $\mathcal{L}_2$ Gain and $\mathcal{H}_\infty$ Gain of a Nonlinear System

(4.6) is locally attractive, and an integral curve $x(\cdot)$ of (4.4) approaches a uniquely defined integral curve $x_{ss}(\cdot)$, passing through the initial state $\pi(w(0))$ provided that the exogenous input $w(t)$ remains in W for all $t \geq 0$ and the initial state $x(0)$ is sufficiently close to the central manifold (4.6).

The *steady-state response* $x_{ss}(t)$ to the periodic input $w(t)$ is thus defined for the nonlinear system (4.4). Since manifold (4.6) is invariant, the integral curve $x_{ss}(\cdot)$ is such that $x_{ss}(t) = \pi(w(t))$ for all t and the steady-state response $x_{ss}(t)$ is T-periodic because the exogenous input $w(t)$ is so. Using the center manifold (4.6), we can now introduce a steady-state response $x_{ss}(t)$ for the state of the nonlinear system (4.1) and a steady-state response $z_{ss}(t)$ for its output variable (4.2). Suppose that (4.1) is locally exponentially stable and is affected by a T-periodic signal $w(t)$ so that both $x_{ss}(t)$ and $z_{ss}(t)$ are periodic functions of the same period T. Evaluating (4.3) on the steady-state response yields

$$\int_0^T \|z_{ss}(t)\|^2 \, dt < \gamma^2 \int_0^T \|w(t)\|^2 \, dt, \tag{4.7}$$

which can be rewritten in the equivalent form

$$\|z_{ss}(\cdot)\|_T < \gamma \|w(\cdot)\|_T, \tag{4.8}$$

given in terms of

$$\|z_{ss}(\cdot)\|_T = \frac{1}{T}\sqrt{\int_0^T \|z_{ss}(t)\|^2 \, dt}, \quad \|w(\cdot)\|_T = \frac{1}{T}\sqrt{\int_0^T \|w(t)\|^2 \, dt}. \tag{4.9}$$

If we define

$$\|T_{zw}\|_\infty = \sup_{w(\cdot)\in\mathcal{W}} \frac{\|z_{ss}(\cdot)\|_T}{\|w(\cdot)\|_T} \tag{4.10}$$

over the class $\mathcal{W}$ of periodic continuous inputs $w(\cdot)$ of any period T subject to $w(t) \in W$ for all t (i.e., for which there exists a steady-state response), it follows that

$$\|T_{zw}\|_\infty \leq \gamma, \tag{4.11}$$

provided that system (4.1), (4.2) has $\mathcal{L}_2$ gain less than γ.

Actually, the quantity $\|T_{zw}\|_\infty$, thus defined, inherits the same features of the $\mathcal{H}_\infty$ norm of a linear system and it is typically referred to as the $\mathcal{H}_\infty$ gain of the nonlinear system (4.1), (4.2) that has become standard in the literature.

4.2 Control Objective

The nonlinear $\mathcal{H}_\infty$-control problem of interest is stated below for a system of the form

$$\dot{x}(t) = f(x(t),t) + g_1(x(t),t)w(t) + g_2(x(t),t)u(t), \qquad (4.12)$$
$$z(t) = h_1(x(t),t) + k_{12}(x(t),t)u(t), \qquad (4.13)$$
$$y(t) = h_2(x(t),t) + k_{21}(x(t),t)w(t), \qquad (4.14)$$

where $x \in \mathbb{R}^n$ is the state vector, $u \in \mathbb{R}^m$ is the control input, $w \in \mathbb{R}^r$ is the unknown disturbance, $z \in \mathbb{R}^l$ is the unknown output to be controlled, and $y \in \mathbb{R}^p$ is the only available measurement on the system. The following assumptions are made throughout:

- the functions $f(x), g_1(x), g_2(x), h_1(x), h_2(x), k_{12}(x), k_{21}(x)$ are of appropriate dimensions and of class C^∞;
- $f(0) = 0, h_1(0) = 0,$ and $h_2(0) = 0$;
-

$$h_1^T(x)k_{12}(x) = 0, \; k_{12}^T(x)k_{12}(x) = I,$$
$$k_{21}(x)g_1^T(x,t) = 0, \; k_{21}(x)k_{21}^T(x) = I. \qquad (4.15)$$

The former assumptions are typical for the nonlinear treatment [67], whereas the latter contains simplifying assumptions, inherited from the linear treatment.

For later use, the notion of an admissible controller is due. Consider a causal dynamic feedback controller of the form

$$\dot{\xi} = \eta(\xi, y),$$
$$u = \theta(\xi) \qquad (4.16)$$

with the internal state $\xi \in \mathbb{R}^s$ and with the C^k-functions η and θ, corresponding to some integer $k \geq 1$ and satisfying $\eta(0,0) = 0, \theta(0) = 0$. Such a controller (4.16) is said to be a *globally (locally) admissible controller* if the closed-loop system (4.12)–(4.16) is globally (uniformly) asymptotically stable when $w \equiv 0$.

The nonlinear $\mathcal{H}_\infty$-control objective that corresponds to the disturbance attenuation level γ is to find, if possible, a globally (locally) admissible dynamic feedback (4.16) such that the $\mathcal{L}_2$ gain of the closed-loop system (4.12)–(4.16) is less than γ (respectively, locally around the origin). Clearly, the control problem thus formulated requires a controller design that guarantees both the internal asymptotic stability of the closed-loop system and its dissipativity with respect to admissible external disturbances.

4.3 Isaacs' Equation and State-Space Solution

The Isaacs equation forms the conceptual foundation of the nonlinear $\mathcal{H}_\infty$ synthesis. We review its heuristic derivation first. Suppose that the inequality

$$V(x(t_1)) - V(x(t_0)) + \int_{t_0}^{t_1} L(x(\tau), w(\tau), \alpha(x(\tau), w(\tau)))d\tau \leq 0, \qquad (4.17)$$

where

$$L(x, w, u) = \|h_1(x) + k_{12}(x)u\|^2 - \gamma^2 \|w\|^2 \qquad (4.18)$$

is satisfied with a positive semidefinite C^1-function $V(x)$ and with a feedback law $\alpha(x, w)$ for all $t_1 \geq t_0$ and for any solution $x(t)$ of the closed-loop system

$$\dot{x} = f(x) + g_1(x)w + g_2(x)\alpha(x, w). \qquad (4.19)$$

Then, letting $t_0 = 0$ and $t_1 = T$ in (4.17) and taking into account that

$$V(x(t_1)) - V(x(t_0)) = V(x(T)) \geq 0 \qquad (4.20)$$

for any trajectory of (4.19), initialized with $x(0) = 0$, we find that

$$\int_0^T L(x(\tau), w(\tau), \alpha(x(\tau), w(\tau)))d\tau \leq 0, \qquad (4.21)$$

thereby yielding (4.3) and thus ensuring that the closed-loop system (4.19) possesses the $\mathcal{L}_2$ *gain less than* γ with respect to output (4.2).

Passing from (4.17) to its infinitesimal version, one arrives at

$$\frac{\partial V}{\partial x}[f(x) + g_1(x)w + g_2(x)\alpha(x, w)] + L(x, w, \alpha(x, w)) \leq 0. \qquad (4.22)$$

The question then arises as to whether there exist a feedback $\alpha(x, w)$ and a positive semidefinite function $V(x)$, satisfying (4.22). To address this question, let us introduce the function

$$H(x, p^T, w, u) = p^T[f(x) + g_1(x)w + g_2(x)u] + L(x, w, u). \qquad (4.23)$$

Since under the simplifying assumptions, made earlier, function (4.18) takes the form

$$L(x, w, u) = h_1^T(x)h_1(x) + u^T u - \gamma^2 w^T w,$$

relation (4.23) determines the quadratic function H in the arguments u and w, which is why it is representable in the truncated Taylor series

$$H(x, p^T, w, u) = H(x, p^T, \alpha_1, \alpha_2) - \gamma^2 \|w - \alpha_1\|^2 + \|u - \alpha_2\|^2, \quad (4.24)$$

where

$$\alpha_1(x, p^T) = \frac{1}{2\gamma^2} g_1^T(x) p, \quad \alpha_2(x, p^T) = -\frac{1}{2} g_2^T(x) p \quad (4.25)$$

are unique extremal points of the quadratic function H such that

$$\frac{\partial H(x, p^T, w, u)}{\partial w} = 0 \quad \text{and} \quad \frac{\partial H(x, p^T, w, u)}{\partial u} = 0 \text{ at } (w, u) = (\alpha_1, \alpha_2).$$

If it happens that α_1 and α_2 solve the equation

$$H(x, V_x, \alpha_1(x, V_x), \alpha_2(x, V_x)) = 0, \quad (4.26)$$

where V_x denotes the Jacobian matrix $\partial V/\partial x$ of V, then (4.24) results in

$$H(x, V_x, w, u) = -\gamma^2 \|w - \alpha_1(x, V_x)\|^2 + \|u - \alpha_2(x, V_x)\|^2, \quad (4.27)$$

and by definition (4.23) of the function H, it follows that the control law

$$u = \alpha_2(x, V_x)) \quad (4.28)$$

renders the infinitesimal version (4.22) of the dissipation inequality (4.17). Using (4.25), we can rewrite Eq. (4.26) in the standard form of Isaacs' equation:

$$V_x f(x) + h_1^T(x) h_1(x) + \frac{1}{4\gamma^2} V_x g_1(x) g_1^T(x) V_x^T - \frac{1}{4} V_x g_2(x) g_2^T(x) V_x^T = 0. \quad (4.29)$$

Isaacs' equation (4.29) is a Hamilton–Jacobi PDE for $V(x)$. The existence of a C^2 positive semidefinite solution $V(x)$ of such an equation is well known (see, e.g., [15]) to be necessary and in the case of $V_x(0) = 0$ (e.g., if V is positive definite) also sufficient for the existence of a special invariant manifold

$$M = \{(x, p) : p = V_x^T\} \quad (4.30)$$

of the Hamiltonian system

$$\frac{dx}{dt} = \left(\frac{\partial H_*(x, p)}{\partial p}\right)^T, \quad \frac{dp}{dt} = -\left(\frac{\partial H_*(x, p)}{\partial x}\right)^T, \quad (4.31)$$

4.3 Isaacs' Equation and State-Space Solution

generated by the Hamiltonian function

$$H_*(x, p) = H(x, p^T, \alpha_1(x, p^T), \alpha_2(x, p^T))), \quad (4.32)$$

where α_1 and α_2 are given by (4.25).

The existence of the invariant manifold (4.30) for the Hamiltonian system (4.29) was studied by Van der Schaft in [121]. Due to [121], such an invariant manifold exists and Isaacs' equation (4.31) is therefore solvable (thereby ensuring that the nonlinear $\mathcal{H}_\infty$-control problem can be locally solved with the disturbance attenuation level γ) if the linear approximation of (4.12)–(4.14) at the equilibrium admits a linear state feedback such that the $\mathcal{H}_\infty$ norm of the resulting linear closed-loop system is less than γ. This result allowed Van der Schaft [122] to develop the nonlinear state-feedback $\mathcal{H}_\infty$ synthesis that was then generalized by Isidori and Astolfi [67] for the nonlinear output-feedback $\mathcal{H}_\infty$ synthesis.

The nonlinear synthesis, proposed in [67], relies on solving two coupled Hamilton–Jacobi–Isaacs PDEs and is based on the following hypotheses:

H1. The pair $\{f, h_1\}$ is *locally detectable*, that is, there exists a neighborhood U of the origin in $\mathbb{R}^n$ such that if $x(t)$ is a solution of $\dot{x} = f(x)$, satisfying $x(0) \in U$, then $h_1(x(t))$ is defined for all $t \geq 0$ and $h_1(x(t)) = 0$ for all $t \geq 0$ implies that $\lim_{t \to \infty} x(t) = 0$.

H2. There exists a smooth positive definite function $V(x)$, locally defined in a neighborhood of the origin $x = 0$, that solves the Hamilton–Jacobi–Isaacs equation (4.29).

H3. There exists an $p \times n$ matrix G such that the equilibrium $\xi = 0$ of the system

$$\dot{\xi} = f(\xi) + g_1(\xi)\alpha_1(\xi) - G h_2(\xi) \quad (4.33)$$

is locally asymptotically stable provided that $\alpha_1(\xi) = \frac{1}{2}\gamma^{-2} g_1^T(x) V_x^T$.

H4. There exists a smooth positive semidefinite function $W(x, \xi)$, locally defined in a neighborhood of the origin in $\mathbb{R}^n \times \mathbb{R}^n$ and such that $W(0, \xi)$ is positive definite, that solves the Hamilton–Jacobi–Isaacs equation

$$\left(\frac{\partial W}{\partial x} \quad \frac{\partial W}{\partial \xi} \right) f_e(x, \xi) + h_e^T(x, \xi) h_e(x, \xi) + \gamma^2 \phi^T(x, \xi) \phi(x, \xi) = 0, \quad (4.34)$$

where

$$f_e^T(x, \xi) = ((f_e^1(x, \xi))^T, (f_e^2(x, \xi))^T),$$
$$f_e^1(x, \xi) = f(x) + g_1(x)\alpha_1(x) + g_2(x)\alpha_2(\xi),$$
$$f_e^2(x, \xi) = f(\xi) + g_1(\xi)\alpha_1(\xi) + g_2(\xi)\alpha_2(\xi) + G(t)(h_2(x) - h_2(\xi)),$$
$$h_e(x, \xi) = \alpha_2(\xi) - \alpha_2(x),$$

$$\phi(x,\xi) = \frac{1}{2\gamma^2} g_e^T(x) \left(\begin{array}{c} \left(\frac{\partial W}{\partial x}\right)^T \\ \left(\frac{\partial W}{\partial \xi}\right)^T \end{array} \right),$$

$$g_e(x) = \left(\begin{array}{c} g_1(x) \\ G k_{21}(x) \end{array} \right),$$

and $\alpha_2(\xi) = -\frac{1}{2} g_2^T(x) V_x^T$.

Theorem 15 ([67]). *Let Hypotheses H1–H4 be satisfied. Then the nonlinear $\mathcal{H}_\infty$-control problem is solved by the output feedback*

$$\dot{\xi} = f(\xi) + g_1(\xi)\alpha_1(\xi) + g_2(\xi)\alpha_2(\xi) + G[y - h_2(\xi)],$$
$$u = \alpha_2(\xi). \tag{4.35}$$

Under appropriate assumptions on the linearized system

$$\dot{x} = Ax + B_1 w + B_2 u,$$
$$z = C_1 x + D_{12} u,$$
$$y = C_2 x + D_{21} w, \tag{4.36}$$

where

$$A = \frac{\partial f}{\partial x}(0), \; B_1 = g_1(0), \; B_2 = g_2(0), \; C_1 = \frac{\partial h_1}{\partial x}(0),$$

$$C_2 = \frac{\partial h_2}{\partial x}(0), \; D_{12} = k_{12}(0), \; D_{21} = k_{21}(0), \tag{4.37}$$

the hypotheses of Theorem 15 can readily be tested by solving the standard $\mathcal{H}_\infty$-control problem for this linearized system, and these hypotheses become not simply sufficient but also necessary for the nonlinear $\mathcal{H}_\infty$-control problem to have a solution. The following result, extracted from [67], allows one to locally solve the nonlinear $\mathcal{H}_\infty$-control problem at the same level of complexity as that in the linear case.

Theorem 16. *Suppose there exists a controller that locally exponentially stabilizes the corresponding closed-loop system (4.12)–(4.14) and renders its $\mathcal{L}_2$ gain less than γ. If the linearized system (4.36) specified with (4.37) is such that (A, B_1) is controllable and (A, C_1) is observable, then Conditions (C1)–(C3) of Theorem 1 are satisfied. If, in addition, the controller*

$$\dot{\xi} = (A + B_1 F_1 + B_2 F_2 - G C_2)\xi + G y,$$
$$u = F_2 \xi, \tag{4.38}$$

4.3 Isaacs' Equation and State-Space Solution

specified with $F_1 = \gamma^{-2} B_1^T P$, $F_2 = -B_2^T P$, $G = QC_2^T$, is either controllable or observable, then Hypotheses H1–H4 hold locally with $G = QC_2^T$, $V(x) = x^T P x$, $W(x, \xi) = \gamma^2 (x - \xi)^T Q^{-1}(x - \xi)$, where P and R are positive definite symmetric solutions of the AREs (1.31) and (1.32), respectively, and $Q = R(I - \gamma^{-2} PR)^{-1}$.

In Part II of this book, the nonlinear $\mathcal{H}_\infty$ synthesis will be extended to nonsmooth systems and its capabilities will be illustrated by applications to mechanical systems with nonsmooth phenomena.

Part II
Nonsmooth $\mathcal{H}_\infty$ Control

The state-space approach, which has been developed for the nonlinear $\mathcal{H}_\infty$ synthesis at the same level of generality as that in the linear case, is now extended to the nonsmooth setting. Both the full information case with perfect state measurements and the incomplete information case with output disturbance-corrupted measurements are studied side by side.

Sufficient conditions for the existence of a global solution of the problem are carried out in terms of an appropriate solvability of two Hamilton–Jacobi–Isaacs partial differential inequalities that arise in the state-feedback and output-injection designs, respectively, and that may not admit continuously differentiable solutions. The present $\mathcal{L}_2$-gain analysis follows the line of reasoning where the corresponding Hamilton–Jacobi–Isaacs expressions are viewed in the sense of Clarke proximal superdifferentials and are required to be negative definite rather than semidefinite. This feature allows one to develop an $\mathcal{H}_\infty$-design procedure in the nonsmooth setting with no a priori imposed stabilizability–detectability conditions on the control system. The resulting controller is associated with specific proximal solutions of the Hamilton–Jacobi–Isaacs partial differential inequalities.

Although the design procedure results in an infinite-dimensional problem, this difficulty is circumvented by solving the problem locally. A local solution is derived by means of a certain perturbation of the DREs, which appear in solving the $\mathcal{H}_\infty$-control problem for the linearized system. By invoking the time-varying counterpart of the strict bounded real lemma, the existence of suitable solutions of the perturbed DREs is established whenever the corresponding unperturbed equations possess uniformly bounded positive semidefinite solutions. Stabilizability and detectability properties of the control system are thus ensured by the existence of the proper solutions of the unperturbed DREs, and hence the proposed synthesis procedure obviates extra work on verifying these properties that might present a formidable problem in the nonlinear case and is definitely so in the nonsmooth case.

Being developed in the general time-varying setting, the nonsmooth synthesis is then specified for periodic and autonomous systems with a focus on the periodic and, respectively, time-invariant controller design. Local output-feedback synthesis is additionally presented over sampled-data measurements. An LMI-based extension of the state-space approach to a class of nonsmooth distributed parameter systems finalizes the present development.

Chapter 5
Elements of Nonsmooth Analysis

A differential construct that applies to nonsmooth functions is useful in general. The proximal supergradient [31] admits a very complete calculus for upper semicontinuous functions and perfectly suits the nonsmooth $\mathcal{L}_2$-gain analysis to be developed in this chapter.

5.1 Semicontinuous Functions and Their Proximal Sub- and Superdifferentials

Let $f(x)$ be a real-valued scalar function with finite values that is defined on $\mathbb{R}^n$. Such a function is said to be *upper semicontinuous* at $y \in \mathbb{R}^n$ iff for all $\varepsilon > 0$, there exists $\delta > 0$ such that $f(x) \leq f(y) + \varepsilon$ for any $x \in B_\delta(y)$. Hereinafter, $B_\delta(y) \subset \mathbb{R}^n$ stands for a ball of radius δ, centered at y.

Lower semicontinuity is complementary to upper semicontinuity in the sense that f is *lower semicontinuous* at y iff $-f$ is upper semicontinuous at y; that is, f is lower semicontinuous at $y \in \mathbb{R}^n$ iff for all $\varepsilon > 0$, there exists $\delta > 0$ such that $f(x) \geq f(y) - \varepsilon$ for any $x \in B_\delta(y)$.

It is customary to say that f is *continuous* at y if it is both lower and upper semicontinuous at y. The function f is said to be *locally Lipschitz continuous around* y iff there exists a ball $B_\delta(y)$ such that $\|f(x_1) - f(x_2)\| \leq L\|x_1 - x_2\|$ for all $x_1, x_2 \in B_\delta(y)$ and some positive constant L. The function f is said to be locally Lipschitz continuous iff it is locally Lipschitz continuous around any $y \in \mathbb{R}^n$.

A vector $\zeta(y) \in \mathbb{R}^n$ is said to be a *proximal supergradient* of a scalar function $f(x)$ at $y \in \mathbb{R}^n$ if there exists some $\sigma(y) \geq 0$ such that

$$f(x) \leq f(y) + \zeta^T(y)(x - y) + \sigma(y)\|x - y\|^2 \tag{5.1}$$

for all x in some neighborhood $U(y) \subset \mathbb{R}^n$ of y.

Complementary to the above is the proximal subgradient concept. A vector $\zeta(y) \in \mathbb{R}^n$ is said to be a *proximal subgradient* of $f(x)$ at $y \in \mathbb{R}^n$ if $-\zeta(y)$ is a proximal supergradient of $-f(x)$ at y; that is, there exists some $\sigma(y) \geq 0$ such that

$$f(x) \geq f(y) + \zeta^T(y)(x-y) - \sigma(y)\|x-y\|^2 \quad (5.2)$$

for all x in some neighborhood $U(y) \subset \mathbb{R}^n$ of y.

The sets $\partial^P f(y)$ and $\partial_P f(y)$ of proximal supergradients and proximal subgradients of f at y are referred to as the *proximal superdifferential* and *proximal subdifferential* of f at y, respectively. These sets are definitely convex, but they may be empty, closed, or open as well as bounded or unbounded.

As a trivial exercise, one can establish that the proximal superdifferential of the function $f_1(x) = |x|$ at 0 in dimension $n = 1$ is empty, whereas that of the function $f_2(x) = -|x|$ is the segment $[-1,1]$, namely, $\partial^P f_1(0) = \emptyset$, $\partial^P f_2(0) = [-1,1]$, and vice versa: The relations $\partial_P f_1(0) = [-1,1]$, $\partial_P f_2(0) = \emptyset$ are established for the proximal subdifferentials at 0 of the functions in question.

The existence of a proximal supergradient ζ at y (and that of a proximal subgradient) relies on the capability of the function f to be approximated in a unilateral manner from above (respectively, from below) by a quadratic function with the contact point $(y, f(y))$ and the slope ζ at that point. Note that if f is differentiable at y, then one has $f'(y) \in \partial_P f(y) \cap \partial^P f(y)$, whereas $f'(y) = \partial_P f(y) = \partial^P f(y)$ provided that f is of class C^2 at y.

It should be noted that if confined to either lower semicontinuous or upper semicontinuous functions, the standard differential calculus possesses its complete counterpart in terms of sub- and superdifferentials, respectively. Details are beyond the present scope and may be found in [31].

5.1.1 Proximal Sub- and Supergradients in a Time-Varying Setting

A natural extension of the super- and subgradients' concepts to time-varying functions $g(x,t)$, defined on $\mathbb{R}^{n+1}$, is as follows.

A vector $\eta(y,\tau) = (\zeta^T(y,\tau), s(y,\tau))^T \in \mathbb{R}^n \times \mathbb{R}$ is a *proximal supergradient* of a scalar function $g(x,t)$ at $(y,\tau) \in \mathbb{R}^n \times \mathbb{R}$ iff there exists some $\kappa(y,\tau) \geq 0$ such that

$$g(x,t) \leq g(y,\tau) + \zeta^T(y,\tau)(x-y) + s(y,\tau)(t-\tau) + \kappa(y,\tau)[\|x-y\|^2 + (t-\tau)^2] \quad (5.3)$$

for all (x,t) in some neighborhood $U(y,\tau) \subset \mathbb{R}^n \times \mathbb{R}^1$ of (y,τ).

For time-varying functions $g(x,t)$, the proximal subgradient concept remains complementary to that of proximal supergradient; it is defined as a proximal supergradient of the function $-g(x,t)$ with the factor -1.

5.2 Viscosity Solutions of First-Order PDEs

Similar to the time-invariant case, the sets $\partial^P g(y,t)$ and $\partial_P g(y,t)$ of proximal supergradients and proximal subgradients of g are referred to as the *proximal superdifferential* and, respectively, *proximal subdifferential* of g at (y,t).

5.2 Viscosity Solutions of First-Order PDEs

Let's consider a first-order partial differential equation (PDE) of the form

$$F(x, V, \nabla V) = 0, \qquad (5.4)$$

where F is a certain scalar function of the arguments $x = (x_1, \ldots, x_n)^T \in \mathbb{R}^n$, $V(\cdot) \in \mathbb{R}^1$, $\nabla V(\cdot) \in \mathbb{R}^n$, and the symbol $\nabla = \left(\frac{\partial}{\partial x_1}, \ldots, \frac{\partial}{\partial x_n}\right)^T$ stands for the nabla operator. It should be pointed out that (5.4) may be viewed in the time-varying setting as well when the argument $(x_1, \ldots, x_n, t)^T \in \mathbb{R}^{n+1}$ is formally substituted into (5.15) for x and the operator nabla reads $\nabla = \left(\frac{\partial}{\partial x_1}, \ldots, \frac{\partial}{\partial x_n}, \frac{\partial}{\partial t}\right)^T$, respectively.

The solvability of the above PDE is, in general, not ensured in the class of smooth functions. In order to define continuous solutions of (5.4), Crandall and Lions introduced the viscosity solution concept in [34]. A continuous function $V(x)$ is said to be a *viscosity solution* of (5.4) iff for all x, whenever W is a smooth function such that $V - W$ possesses a local minimum at x, one has

$$F(x, V, \nabla W) \geq 0, \qquad (5.5)$$

and whenever W is a smooth function such that $V - W$ possesses a local maximum at x, one has

$$F(x, V, \nabla W) \leq 0. \qquad (5.6)$$

The set $\partial^D V(x)$ of values $\nabla W(x)$, where W is a smooth function such that $V - W$ possesses a local maximum at x, is referred to as the *Dini (or viscosity) superdifferential* of V at x. The corresponding concept with minimum instead of maximum defines the *Dini (or viscosity) subdifferential* $\partial_D V(x)$. This terminology is due to the fact (see [31] for details) that in the superdifferential case, one has $\zeta \in \partial^D V(x)$ iff the (upper) Dini superderivative

$$D^+ V(x; v) = \limsup_{\mu \to v;\, \tau \to 0+} \frac{1}{\tau}[V(x + \tau\mu) - V(x)] \qquad (5.7)$$

satisfies

$$D^+ V(x; v) \leq \zeta^T v \text{ for any } v \in \mathbb{R}^n. \qquad (5.8)$$

In the subdifferential case, one has $\zeta \in \partial_D V(x)$ iff the (lower) Dini subderivative

$$D_+ V(x; v) = \liminf_{\mu \to v;\, \tau \to 0+} \frac{1}{\tau}[V(x + \tau\mu) - V(x)] \qquad (5.9)$$

satisfies

$$D_+V(x;v) \geq \zeta^T v \text{ for any } v \in \mathbb{R}^n. \tag{5.10}$$

In terms of Dini sub- and superdifferentials, a viscosity solution of (5.4) is redefined as a continuous function $V(x)$ that satisfies the following inequalities:

$$F(x, V, \partial_D V(x)) \geq 0, \tag{5.11}$$

$$F(x, V, \partial^D V(x)) \leq 0 \tag{5.12}$$

for all $x \in \mathbb{R}^n$.

It is of interest to note that the proximal sub- and supergradients correspond to specific linear and quadratic functions $W(x)$ that appear in the viscosity solution definition. Indeed, given $\zeta \in \partial_P V(x)$, the function $V(y) - W(y)$ with a particular $W(y) = \zeta^T y - \sigma \|y - x\|^2$ and some positive constant σ possesses a local minimum at x and ∇W at x equals ζ. Hence, the inequality $F(x, V(x), \zeta) \geq 0$ holds for any $\zeta \in \partial_P V(x)$, thereby yielding the multivalued inequality

$$F(x, V, \partial_P V(x)) \geq 0 \text{ for all } x \in \mathbb{R}^n. \tag{5.13}$$

Similarly, we obtain the multivalued inequality

$$F(x, V, \partial^P V(x)) \leq 0 \text{ for all } x \in \mathbb{R}^n. \tag{5.14}$$

Thus, a viscosity solution of (5.4) necessarily satisfies its proximal bilateral version (5.13) and (5.14). Under appropriate assumptions, the sufficiency of (5.13) and (5.14) for V to be a viscosity solution is established [31] by applying Subbotin's theorem [117], according to which any Dini subgradient can be, roughly speaking, approximated by a proximal subgradient. Numerical methods of seeking viscosity solutions of first-order PDEs, associated with the nonlinear $\mathcal{H}_\infty$ synthesis [14], are systematically studied in [80].

Viscosity solutions are just a way of defining nonsmooth solutions of the first-order PDE (5.4) through bilateral approximations (5.5) and (5.6). Alternative solution concepts may be found in [31]. Apparently, while one deals with the corresponding partial differential inequality

$$F(x, V, \nabla V) \leq 0, \tag{5.15}$$

only the unilateral approximation (5.14), which uses the proximal superdifferential, is relevant. With this in mind, while developing the $\mathcal{L}_2$-gain analysis of nonsmooth systems, one next brings into play proximal solutions of a particular form of (5.14), referred to as a Hamilton–Jacobi inequality.

5.3 Nonsmooth $\mathcal{L}_2$-Gain Analysis

The $\mathcal{L}_2$-gain analysis, presented here, is based on the game-theoretic approach from [14] and extends the results from [67, 122]—where investigations were confined to smooth autonomous systems—toward locally Lipschitz continuous nonautonomous systems.

5.3.1 Basic Assumptions and Definitions

The $\mathcal{L}_2$-gain analysis is developed for a nonautonomous system of the form

$$\dot{x} = \varphi(x,t) + \psi(x,t)w(t) \tag{5.16}$$

and is made with respect to the output

$$z(t) = h(x(t),t). \tag{5.17}$$

Hereinafter, $x \in \mathbb{R}^n$ is the state vector, $t \in \mathbb{R}^1$ is the time variable, $w \in \mathbb{R}^r$ is the unknown disturbance, $\varphi(\cdot,\cdot) \in \mathbb{R}^n$ and $h(\cdot,\cdot) \in \mathbb{R}^p$ are vector functions, and $\psi(\cdot,\cdot) \in \mathbb{R}^{n \times r}$ is a matrix function. The following *assumptions* are imposed on the system:

1. The functions $\varphi(x,t)$, $\psi(x,t)$, and $h(x,t)$ are piecewise-continuous in t for all x and locally Lipschitz continuous in x for almost all t.
2. $\varphi(0,t) = 0$ and $h(0,t) = 0$ for almost all t.

Although the underlying system is admitted to be nonsmooth, however, once it is affected by a locally integrable (particularly, piecewise-continuous) disturbance $w(t)$, the state equation (5.16) is guaranteed, by Assumption 1, to be well posed. In addition, Assumption 2 ensures that the origin is an equilibrium point of the nominal (i.e., disturbance-free) system

$$\dot{x} = \varphi(x,t). \tag{5.18}$$

Given a real number $\gamma > 0$, further referred to as a disturbance attenuation level, system (5.16) is said to have *$\mathcal{L}_2$ gain less than γ* with respect to output (5.17) [or, simply, system (5.16), (5.17) has $\mathcal{L}_2$-gain less than γ] if the response z, resulting from w for initial state $x(t_0) = 0$, satisfies

$$\int_{t_0}^{t_1} \|z(t)\|^2 \, dt < \gamma^2 \int_{t_0}^{t_1} \|w(t)\|^2 \, dt \tag{5.19}$$

for all $t_1 > t_0$ and all piecewise-continuous functions $w(t)$.

Respectively, system (5.16), (5.17) is said to have $\mathcal{L}_2$ *gain less than γ, locally around the origin*, if there exists a neighborhood U of the origin such that inequality (5.19) is satisfied for all $t_1 > t_0$ and all piecewise-continuous functions $w(t)$ for which the state trajectory of the closed-loop system starting from the initial point $x(t_0) = 0$ remains in U for all $t \in [t_0, t_1]$.

5.3.2 Instrumental Lemmas

Technical lemmas are now presented to be used in the subsequent nonsmooth $\mathcal{L}_2$-gain analysis. A standard notation

$$DV(x, t; v, 1) = \lim_{\tau \to 0} \frac{V(x + \tau v, t + \tau) - V(x, t)}{\tau} \quad (5.20)$$

stands throughout for a Dini derivative (if any) of a scalar function $V(x, t)$, computed in the direction $(v^T, 1)^T \in \mathbb{R}^{n+1}$ at $(x, t) \in \mathbb{R}^{n+1}$.

Lemma 8. *Let $x(\cdot) \in \mathbb{R}^n$ be an absolutely continuous function of the time variable t and let $V(x, t)$ be a scalar locally Lipschitz continuous function around any $(x, t) \in \mathbb{R}^{n+1}$. Then the composite function $V(x(t), t)$ is absolutely continuous and its time derivative is given by*

$$\frac{d}{dt} V(x(t), t) = DV(x(t), t; \dot{x}(t), 1) \quad (5.21)$$

almost everywhere. Furthermore,

$$DV(x(t), t; \dot{x}(t), 1) \leq \frac{\partial V}{\partial t} + \frac{\partial V}{\partial x} \dot{x}(t) \quad (5.22)$$

for almost all t and for all supergradients $\left(\frac{\partial V}{\partial x}, \frac{\partial V}{\partial t}\right)^T \in \partial^P V(x, t)$, if any.

Proof. Given time instant t, let us employ the absolute continuity of $x(t)$ to choose an interval $I = (t - \delta, t + \delta)$ with some $\delta > 0$ such that $x(\tau) \in B_\delta(x(t))$ for all $\tau \in I$. Then, due to the local Lipschitz continuity of $V(x, t)$, the Lipschitz inequality

$$|V(x, s) - V(\xi, \tau)| \leq L [\|x - \xi\| + |t - \tau|]$$

holds for all $x, \xi \in B_{\delta'}(x(t))$, for all $s, \tau \in I$, and some $L > 0$, $\delta' \in (0, \delta)$. Let's now fix an arbitrary $\varepsilon > 0$, and let's choose, by virtue of the absolute continuity of $x(t)$, a positive $\delta'' < \min\left\{\delta', \frac{\varepsilon}{2L}\right\}$ such that the inequality

$$\Sigma_i \|x(t_i) - x(\tau_i)\| < \frac{\varepsilon}{2L}$$

5.3 Nonsmooth $\mathcal{L}_2$-Gain Analysis

holds whenever a finite sequence of pairwise-disjoint subintervals (t_i, τ_i) satisfies $\Sigma_i |t_i - \tau_i| < \delta''$. It follows that

$$\Sigma_i |V(x(t_i), t_i) - V(x(\tau_i), \tau_i)| \le L\Sigma_i [\|x(t_i) - x(\tau_i)\| + |t_i - \tau_i|] <$$
$$L\left(\frac{\varepsilon}{2L} + \delta''\right) < L\left(\frac{\varepsilon}{2L} + \frac{\varepsilon}{2L}\right) = \varepsilon.$$

Hence, the composite function $V(x(t), t)$ is absolutely continuous and differentiable almost everywhere.

In order to demonstrate the validity of (5.21), we derive

$$\frac{d}{dt}V(x(t), t) = \lim_{\tau \to 0} \frac{V(x(t+\tau), t+\tau) - V(x(t), t)}{\tau}$$
$$= \lim_{\tau \to 0} \frac{V(x(t) + \tau\dot{x}(t), t+\tau) - V(x(t), t)}{\tau}$$
$$+ \lim_{\tau \to 0} \frac{V(x(t+\tau), t+\tau) - V(x(t) + \tau\dot{x}(t), t+\tau)}{\tau}$$
$$= \lim_{\tau \to 0} \frac{V(x(t) + \tau\dot{x}(t), t+\tau) - V(x(t), t)}{\tau} = D(x(t), t; \dot{x}(t), 1) \quad (5.23)$$

at any time instant t, where the derivatives $\dot{x}(t)$ and $dV(x(t), t)/dt$ exist. To reproduce (5.23), noting

$$\lim_{\tau \to 0} \frac{V(x(t+\tau), t+\tau) - V(x(t) + \tau\dot{x}(t), t+\tau)}{\tau} = 0, \quad (5.24)$$

is sufficient, because the function $V(x, t)$ satisfies the Lipschitz condition around (x, t) and $x(t + \tau) = x(t) + \tau\dot{x}(t) + o(\tau)$, where $o(\tau)$ is such that $\lim_{\tau \to 0} \frac{o(\tau)}{\tau} = 0$.

It remains to justify (5.22). It follows from definition (5.3) of the proximal supergradient that

$$\frac{V(x + \tau\dot{x}(t), t + \tau) - V(x, t)}{\tau} \le \frac{\partial V}{\partial t} + \frac{\partial V}{\partial x}\dot{x}(t) + \kappa\tau[\|\dot{x}(t)\|^2 + 1] \quad (5.25)$$

for all supergradients $\left(\frac{\partial V}{\partial x}, \frac{\partial V}{\partial t}\right)^T \in \partial^P V(x, t)$ and some constant $\kappa \ge 0$ [possibly dependent on $x(t)$ and t]. By taking a limit at (5.25) as $\tau \to 0$ for any t where the Dini derivative exists (5.21), the validity of (5.22) is then established. The lemma is thus proved.

For later use, recall the following.

A continuous scalar function $v(x)$ is *positive definite* iff $v(0) = 0$ and $v(x) > 0$ for all $x \ne 0$. It is *radially unbounded* iff $\lim v(x) = \infty$ as $\|x\| \to \infty$. A continuous scalar function $V(x, t)$ is *positive definite* iff $V(x, t) \ge v(x)$ for all $(x, t) \in \mathbb{R}^{n+1}$ and some positive definite function $v(x)$, *radially unbounded* iff $v(x)$ is so, and *decrescent* iff $V(x,t) \leq v_1(x)$ for all $(x,t) \in \mathbb{R}^{n+1}$ and some positive definite function $v_1(x)$. It is *positive semidefinite* iff $V(x,t) \geq 0$ for all $(x,t) \in \mathbb{R}^{n+1}$ and *negative definite (semidefinite)* iff $-V(x,t)$ is positive definite (semidefinite).

A locally Lipschitz continuous positive definite decrescent function $V(x,t)$ is called a *Lyapunov function* of system (5.18) if the time derivative of the composite function $V(x(t),t)$, computed on the solutions of (5.18) according to (5.21), is such that $dV(x(t),t)/dt \leq 0$ for almost all t. If, in addition, $dV(x(t),t)/dt \leq -v(x(t))$ for almost all t and for some positive definite $v(x)$, the function $V(x(t),t)$ is called a strict Lyapunov function of (5.18).

Lemma 9. *Let system (5.18) possess a Lyapunov function $V(x,t)$. Then system (5.18) is uniformly stable. If, in addition, the function $V(x,t)$ is a strict Lyapunov function (and radially unbounded), then system (5.18) is (globally) uniformly asymptotically stable.*

Proof. By Lemma 8, the composite function $V(x(t),t)$, computed on the solutions of (5.18), is absolutely continuous and its time derivative $dV(x(t),t)/dt$ is given by (5.21) almost everywhere. The rest of the proof is identical to its standard counterpart (see, e.g., [71]) as if $V(x(t),t)$ were a smooth Lyapunov function, except for some relations that hold almost everywhere rather than everywhere. The details can be found in [104] and are left to the reader.

5.3.3 Hamilton–Jacobi Inequality and Its Proximal Solutions

We will subsequently analyze system (5.16), (5.17) under the following hypothesis:

H. There exists a locally Lipschitz continuous positive definite decrescent, radially unbounded proximal solution of the Hamilton–Jacobi inequality

$$\frac{\partial V}{\partial t} + \frac{\partial V}{\partial x}\varphi(x,t) + \frac{1}{4\gamma^2}\frac{\partial V}{\partial x}\psi(x,t)\psi^T(x,t)\left(\frac{\partial V}{\partial x}\right)^T + h^T(x,t)h(x,t) \leq -v(x) \tag{5.26}$$

under some positive γ and some positive definite function $v(x)$.

In a particular case of the first-order partial differential inequality (5.14), specified in the time-varying setting (5.26), a proximal solution is defined as follows.

A locally Lipschitz continuous function $V(x,t)$ is said to be a *proximal solution* of the partial differential inequality (5.26) iff its proximal superdifferential $\partial^P V(x,t)$ is everywhere nonempty and (5.26) holds with $V(x,t)$ for all $x \in \mathbb{R}^n$, for almost all $t \in \mathbb{R}$, and for all proximal supergradients $\left(\frac{\partial V}{\partial x}, \frac{\partial V}{\partial t}\right)^T \in \partial^P V(x,t)$.

5.3.4 Global Analysis

The following result presents sufficient conditions of the nonsmooth system (5.16), (5.17) to be internally asymptotically stable and to possess $\mathcal{L}_2$ gain less than γ.

Theorem 17. *Let Assumptions 1 and 2 be in force, and let Hypothesis H be satisfied. Then the nominal system (5.18) is globally asymptotically stable, whereas its disturbed version (5.16) has $\mathcal{L}_2$ gain less than γ with respect to output (5.17).*

Proof. It is clear that Lemma 8 is applicable to a proximal solution $V(x,t)$ of the Hamilton–Jacobi inequality (5.26) viewed on the solutions $x(t)$ of the disturbance-free system (5.18). Then relations (5.21), (5.22), (5.26), coupled together, result in

$$\frac{d}{dt}V(x(t),t) = DV(x(t),t;\dot{x}(t),1) \le \frac{\partial V}{\partial t} + \frac{\partial V}{\partial x}\dot{x}(t)$$

$$= \frac{\partial V}{\partial t} + \frac{\partial V}{\partial x}\varphi(x(t),t) \le -\nu(x(t)). \qquad (5.27)$$

If we take into account that (5.27) holds almost everywhere, Hypothesis 1 thus ensures that $V(x,t)$ is a strict decrescent radially unbounded Lyapunov function of the nominal system (5.18). By Lemma 9, system (5.18) is globally asymptotically stable.

It remains to show that the disturbed system (5.16) has $\mathcal{L}_2$ gain less than γ with respect to output (5.17). For this purpose, we'll introduce the multivalued function

$$H(x,w,t) = \frac{\partial V(x,t)}{\partial t} + \frac{\partial V(x,t)}{\partial x}[\phi(x,t) + \psi(x,t)w]$$

$$+ h^T(x,t)h(x,t) - \gamma^2 w^T w, \qquad (5.28)$$

where $\left(\frac{\partial V}{\partial x}, \frac{\partial V}{\partial t}\right)^T \in \partial^P V(x,t)$. Clearly, the multivalued function (5.17) is quadratic in w. Then

$$\frac{\partial H(x,w,t)}{\partial w}\bigg|_{w=\alpha(x,t)} = \frac{\partial V(x,t)}{\partial x}\psi(x,t) - 2\gamma^2\alpha^T(x,t) = 0 \qquad (5.29)$$

for $\alpha(x,t) = \frac{1}{2\gamma^2}\psi^T(x,t)\left(\frac{\partial V(x,t)}{\partial x}\right)^T$ and $\left(\frac{\partial V}{\partial x}, \frac{\partial V}{\partial t}\right)^T \in \partial^P V(x,t)$. Expanding the quadratic function $H(x,w,t)$ in Taylor series, we derive that

$$H(x,w,t) = H(x,\alpha(x,t),t) - \gamma^2 \|w - \alpha(x,t)\|^2, \qquad (5.30)$$

where $H(x,\alpha(x,t),t) \le -\nu(x)$ due to (5.26). Hence,

$$H(x,w,t) \le -\gamma^2 \|w - \alpha(x,t)\|^2 - \nu(x), \qquad (5.31)$$

and employing (5.28) and (5.31), we arrive at

$$\frac{\partial V(x,t)}{\partial t} + \frac{\partial V(x,t)}{\partial x}[\phi(x,t) + \psi(x,t)w]$$
$$\leq -\gamma^2 \|w - \alpha(x,t)\|^2 - v(x) - \|h(x,t)\|^2 + \gamma^2 \|w\|^2. \qquad (5.32)$$

Applying Lemma 8 and taking (5.32) into account, we can estimate the time derivative of the solution $V(x,t)$ of the Hamilton–Jacobi inequality (5.26) on the trajectories of (5.16), (5.17) as follows:

$$\frac{d}{dt} V(x(t),t) \leq -\gamma^2 \|w(t) - \alpha(x(t),t)\|^2 - v(x(t)) - \|z(t)\|^2 + \gamma^2 \|w(t)\|^2. \quad (5.33)$$

As a matter of fact, the latter inequality ensures that

$$\int_{t_0}^{t_1} (\gamma^2 \|w(t)\|^2 - \|z(t)\|^2) dt \geq V(x(t_1), t_1) - V(x(t_0), t_0)$$
$$+ \gamma^2 \int_{t_0}^{t_1} [\|w(t) - \alpha(x(t),t)\|^2 + v(x(t))] dt > 0 \qquad (5.34)$$

for any trajectory of (5.16), (5.17), initialized with $x(t_0) = 0$. Thus, inequality (5.19) is established and the proof of Theorem 17 is completed.

5.3.5 Local Analysis

In our local analysis, the function $\varphi(x,t) = \varphi_1(x,t) + \varphi_2(x,t)$ is decomposed into two components; for technical reasons, nonsmooth nonlinearities are assumed to be absorbed into the term $\varphi_2(x,t)$ only, whereas $\varphi_1(x,t)$ is assumed to be sufficiently smooth. The following assumptions are thus additionally made:

3. The function $\varphi(x,t)$ admits representation $\varphi(x,t) = \varphi_1(x,t) + \varphi_2(x,t)$; in some neighborhood $U(0)$ of the origin $x = 0$, the functions $\varphi_1(x,t)$, $\psi(x,t)$, and $h(x,t)$ are uniformly bounded in $t \in \mathbb{R}$ and twice continuously differentiable in $x \in U(0)$, whereas their first- and second-order state derivatives are piecewise-continuous and uniformly bounded in $t \in \mathbb{R}$ for all $x \in U(0)$.
4. Given t, the vector $\zeta = 0$ is a proximal time-uniform supergradient of the components $\varphi_{2i}(x,t)$, $i = 1,\ldots,n$, of the function $\varphi_2(x,t)$ at $x = 0$ for almost all t; that is,

$$\varphi_{2i}(x,t) \leq \sigma \|x\|^2 \qquad (5.35)$$

for some $\sigma > 0$, almost all $t \in \mathbb{R}$, and all x in some neighborhood $U(0)$ of the origin.

5.3 Nonsmooth $\mathcal{L}_2$-Gain Analysis

Assumptions 1–4, coupled together, allow one to linearize the corresponding Hamilton–Jacobi inequality, thereby yielding a local $\mathcal{L}_2$-gain analysis. It should be pointed out, however, that estimate (5.35) is unilateral and does not imply that the nonsmooth term $\varphi_2(x)$ becomes negligible in the subsequent local analysis. For instance, (5.35) holds with the scalar function $\varphi_2(x) = -|x|$, whose influence should be taken into account while locally analyzing the system behavior around its equilibrium point. Thus, the local analysis to be developed has to operate with unilateral estimates, and hence it does not represent a straightforward extension of the standard nonlinear $\mathcal{L}_2$ techniques [67, 122].

The local $\mathcal{L}_2$-gain analysis of the underlying system (5.16), (5.17) relies on that of its linearization

$$\dot{x} = A(t)x + B(t)w, \quad z = C(t)x, \tag{5.36}$$

where the matrix functions

$$A(t) = \frac{\partial \varphi_1}{\partial x}(0,t), \quad B(t) = \psi(0,t), \quad C(t) = \frac{\partial h}{\partial x}(0,t) \tag{5.37}$$

are piecewise-continuous and uniformly bounded in $t \in \mathbb{R}$ due to the above assumptions that have been taken into account. Applied to system (5.36), the Hamilton–Jacobi inequality (5.26), rewritten in terms of the quadratic functions $V(x,t) = x^T P_\varepsilon(t)x$ and $v(x) = \varepsilon \|x\|^2$, boils down to the differential Riccati equation

$$\dot{P}_\varepsilon + P_\varepsilon(t)A(t) + A^T(t)P_\varepsilon(t) + C^T(t)C(t) + \frac{1}{\gamma^2}P_\varepsilon(t)B(t)B^T(t)P_\varepsilon(t) + \varepsilon I = 0 \tag{5.38}$$

with the identical matrix I and some $\varepsilon > 0$. By the time-varying strict bounded real Lemma 6, this inequality possesses a uniformly bounded positive definite symmetric solution $P_\varepsilon(t)$ under sufficiently small $\varepsilon > 0$ whenever

C. the corresponding Riccati equation

$$\dot{P} + P(t)A(t) + A^T(t)P(t) + C^T(t)C(t) + \frac{1}{\gamma^2}P(t)B(t)B^T(t)P(t) = 0 \tag{5.39}$$

has a uniformly bounded positive semidefinite symmetric solution $P(t)$ such that system

$$\dot{x} = [A + \gamma^{-2}BB^T P](t)x(t) \tag{5.40}$$

is exponentially stable.

In what follows, Eq. (5.38) is utilized for the local $\mathcal{L}_2$-gain analysis of system (5.16), (5.17).

Theorem 18. *Consider the nominal system (5.18) and its disturbed version (5.16), (5.17) with Assumptions 1–4. Let Condition C be satisfied, and let $P_\varepsilon(t)$ be a uniformly bounded positive definite solution of (5.38) under some $\varepsilon > 0$. Then Hypothesis H holds locally around the equilibrium $x = 0$ with the quadratic functions*

$$V(x,t) = x^T P_\varepsilon(t) x, \qquad (5.41)$$

$$v(x) = \frac{\varepsilon}{2} \|x\|^2, \qquad (5.42)$$

and the nominal system (5.18) is uniformly asymptotically stable, whereas the disturbed system (5.16), (5.17) has $\mathcal{L}_2$ gain less than γ locally around $x = 0$.

Proof. First, note that the function $V(x,t) = x^T P_\varepsilon(t) x$ is differentiable, positive definite, and decrescent due to the conditions of the theorem imposed on the matrix function $P_\varepsilon(t)$. Next, we'll demonstrate that this function satisfies the Hamilton–Jacobi inequality (5.26) in a neighborhood of the origin $x = 0$ for almost all $t \in \mathbb{R}$ provided that $v(x) = \frac{\varepsilon}{2} \|x\|^2$. Indeed,

$$\frac{\partial V}{\partial t} = x^T \dot{P}_\varepsilon(t) x, \qquad (5.43)$$

$$\frac{\partial V}{\partial x} \varphi_1(x,t) = x^T [P_\varepsilon A + A^T P_\varepsilon](t) x + o_t(\|x\|^2), \qquad (5.44)$$

$$\frac{\partial V}{\partial x} \varphi_2(x,t) \leq \sigma^0 \|x\|^3, \qquad (5.45)$$

$$\frac{1}{4\gamma^2} \frac{\partial V}{\partial x} \psi(x,t) \psi^T(x,t) \left(\frac{\partial V}{\partial x}\right)^T = \frac{1}{\gamma^2} x^T P_\varepsilon B(t) B^T(t) P_\varepsilon x + o_t(\|x\|^2), \qquad (5.46)$$

$$h^T(x,t) h(x,t) = x^T C^T(t) C(t) x + o_t(\|x\|^2), \qquad (5.47)$$

where $\sigma^0 = 2\sigma \sup_{t \in \mathbb{R}} \|P_\varepsilon(t)\|$, and by virtue of Assumption 3, $\frac{o_t(\|x\|^2)}{\|x\|^2} \to 0$ uniformly in t as $\|x\|^2 \to 0$. Then, due to (5.38), we have

$$\frac{\partial V}{\partial t} + \frac{\partial V}{\partial x} [\varphi_1(x,t) + \varphi_2(x,t)] +$$

$$\frac{1}{4\gamma^2} \frac{\partial V}{\partial x} \psi(x,t) \psi^T(x,t) \left(\frac{\partial V}{\partial x}\right)^T + h^T(x,t) h(x,t) \leq$$

$$x^T \left[\dot{P}_\varepsilon + P_\varepsilon A + A^T P_\varepsilon + \frac{1}{\gamma^2} P_\varepsilon B B^T P_\varepsilon + C^T C\right](t) x +$$

$$\sigma^0 \|x\|^3 + o_t(\|x\|^2) \leq -\varepsilon \|x\|^2 + o_t(\|x\|^2) \leq -\frac{\varepsilon}{2} \|x\|^2 \qquad (5.48)$$

for almost all $t \in \mathbb{R}$ and $\|x\|$ sufficiently small. Hence, Hypothesis H holds locally with $V(x,t)$ and $v(x)$, given by (5.41) and (5.42), respectively.

Finally, following the line of reasoning used in the proof of Theorem 17, we conclude that the nominal system (5.18) is uniformly asymptotically stable, whereas its disturbed version (5.16), (5.17) has $\mathcal{L}_2$ gain less than γ locally around $x = 0$. Theorem 18 is proved.

5.3.6 Periodic and Autonomous Cases

In the periodic case or, particularly, in the autonomous case, where the functions $\varphi(x,t), \psi(x,t), h(x,t)$ are time-periodic of a period $T > 0$ or, respectively, time-independent, the differential Riccati equations (DREs) (5.38), (5.39) appear with the time-periodic matrices $A(t), B(t), C(t)$ or degenerate to the corresponding algebraic Riccati equations (AREs) obtained by setting $\dot{P} = 0, \dot{P}_\varepsilon = 0$. In the periodic case, Condition C is modified to

C'. Provided that the matrix functions $A(t), B(t), C(t)$ are time-periodic of the period $T > 0$, the DRE (5.39) has a periodic positive semidefinite symmetric solution $P(t)$ of the period T such that system (5.40) is exponentially stable.

In the autonomous case, Condition C is simplified to

C''. Provided that the matrix functions $A(t), B(t), C(t)$ are time-invariant, the ARE

$$PA + A^T P + C^T C + \frac{1}{\gamma^2} PBB^T P = 0 \tag{5.49}$$

has a positive semidefinite symmetric solution P such that the matrix $A + \frac{1}{\gamma^2} BB^T P$ has all eigenvalues with negative real part.

According to the periodic strict bounded real Lemma 7, Condition C' ensures that there exists a positive constant ε_0 such that the perturbed periodic Riccati equation (5.39) has a unique periodic positive definite symmetric solution $P_\varepsilon(t)$ for each $\varepsilon \in (0, \varepsilon_0)$. In turn, according to the strict bounded real lemma [5] (cf. the periodic strict bounded real Lemma 7 that represents the time-invariant case, in particular), Condition C'' ensures that the perturbed ARE

$$P_\varepsilon A + A^T P_\varepsilon + C^T C + \frac{1}{\gamma^2} P_\varepsilon BB^T P_\varepsilon + \varepsilon I = 0 \tag{5.50}$$

has a unique positive definite symmetric solution P_ε, also for each $\varepsilon \in (0, \varepsilon_0)$.

Thus, we arrive at time-periodic and time-invariant counterparts of Theorem 18.

Theorem 19. *Consider the nominal system (5.18) and its disturbed version (5.16), (5.17) with Assumptions 1–4 and with the time-periodic functions $\varphi(x,t), \psi(x,t), h(x,t)$ of a period $T > 0$. Let Condition C' be satisfied and let $P_\varepsilon(t)$ be a periodic positive definite symmetric solution of the periodic DRE (5.38) under some $\varepsilon > 0$. Then Hypothesis H holds locally around the equilibrium $x = 0$ with the time-periodic quadratic function $V(x,t) = x^T P_\varepsilon(t) x$ and quadratic function $v(x) = \frac{\varepsilon}{2} \|x\|^2$. Furthermore, the nominal periodic system (5.18) is uniformly asymptotically stable, whereas its disturbed version (5.16), (5.17) has $\mathcal{L}_2$ gain less than γ locally around $x = 0$.*

Theorem 20. *Consider the nominal system (5.18) and its disturbed version (5.16), (5.17) with Assumptions 1–4 and with the time-invariant functions $\varphi(x,t) = \varphi(x), \psi(x,t) = \psi(x), h(x,t) = h(x)$. Let Condition C'' be satisfied and let P_ε be a positive definite solution of the perturbed ARE (5.50) under some $\varepsilon > 0$. Then Hypothesis H holds locally around the equilibrium $x = 0$ with the time-invariant quadratic functions $V(x) = x^T P_\varepsilon x$ and $v(x) = \frac{\varepsilon}{2} \|x\|^2$. Furthermore, the autonomous nominal system (5.18) is asymptotically stable, whereas its disturbed version (5.16), (5.17) has $\mathcal{L}_2$ gain less than γ locally around $x = 0$.*

Proof of Theorem 19 and that of Theorem 20 are nearly the same as that of Theorem 18, and they are therefore omitted.

The above results, Theorems 17–20, form a basis of the subsequent $\mathcal{H}_\infty$ synthesis of nonsmooth systems in the time-varying, time-periodic, and time-invariant settings.

Chapter 6
Synthesis of Nonsmooth Systems

The nonsmooth $\mathcal{L}_2$-gain analysis developed in Chap. 5 forms a basis for the subsequent $\mathcal{H}_\infty$ synthesis of nonsmooth systems affected by external disturbances. Such a synthesis is first developed via continuous-time measurement feedback and then via sampled-data measurement feedback. Time-varying systems and their periodic and time-invariant counterparts are synthesized side by side in this chapter.

6.1 Synthesis of Time-Varying Systems over Continuous-Time Measurements

The present study focuses on a nonautonomous nonlinear system of the form

$$\dot{x}(t) = f(x(t), t) + g_1(x(t), t)w(t) + g_2(x(t), t)u(t), \tag{6.1}$$

$$z(t) = h_1(x(t), t) + k_{12}(x(t), t)u(t), \tag{6.2}$$

$$y(t) = h_2(x(t), t) + k_{21}(x(t), t)w(t), \tag{6.3}$$

where $x \in \mathbb{R}^n$ is the state vector, $t \in \mathbb{R}$ is the time variable, $u \in \mathbb{R}^m$ is the control input, $w \in \mathbb{R}^r$ is the unknown disturbance, $z \in \mathbb{R}^l$ is the unknown output to be controlled, and $y \in \mathbb{R}^p$ is the only available measurement on the system. In a particular case where the full information $y = x$ is available on the system, the state-feedback synthesis of system (6.1), (6.2) is of interest.

6.1.1 Basic Assumptions and Problem Statement

For the underlying system, the following assumptions are made throughout.

A1. The functions $f(x,t)$, $g_1(x,t)$, $g_2(x,t)$, $h_1(x,t)$, $h_2(x,t)$, $k_{12}(x,t)$, and $k_{21}(x,t)$ are piecewise-continuous in t for all x and locally Lipschitz continuous in x for almost all t.

A2. $f(0,t) = 0$, $h_1(0,t) = 0$, and $h_2(0,t) = 0$ for almost all t.

A3.
$$h_1^T(x,t)k_{12}(x,t) = 0, \quad k_{12}^T(x,t)k_{12}(x,t) = I,$$
$$k_{21}(x,t)g_1^T(x,t) = 0, \quad k_{21}(x,t)k_{21}^T(x,t) = I. \tag{6.4}$$

These assumptions are made for technical reasons. Assumption A1 guarantees the well-posedness of the above dynamic system, while being enforced by integrable exogenous inputs. Along with this, Assumption A1 admits nonsmooth nonlinearities. Assumption A2 ensures that the origin is an equilibrium point of the nondriven ($u = 0$) disturbance-free ($w = 0$) dynamic system (6.1)–(6.3). Assumption A3 is a simplifying assumption inherited from the standard $\mathcal{H}_\infty$-control problem.

In the full information case, a static state-feedback controller

$$u = K(x,t) \tag{6.5}$$

is said to be a globally (locally) *admissible controller* if the closed-loop system (6.1), (6.5) is globally (uniformly) asymptotically stable when $w = 0$. In turn, a causal dynamic output-feedback compensator

$$u = \mathcal{K}(y,t), \tag{6.6}$$

with internal state $\xi \in \mathbb{R}^s$, is said to be a globally (locally) *admissible controller* under the available measurement (6.3) if the closed-loop system (6.1), (6.6) is globally (uniformly) asymptotically stable when $w = 0$.

For convenience of the reader, recall that given a disturbance attenuation level $\gamma > 0$, system (6.1), (6.2), (6.5) [system (6.1)–(6.3), (6.6)] is said to have $\mathcal{L}_2$ gain less than γ if the response z, resulting from w for initial state $x(t_0) = 0$ [and $\xi(t_0) = 0$], satisfies

$$\int_{t_0}^{t_1} \|z(t)\|^2 \, dt < \gamma^2 \int_{t_0}^{t_1} \|w(t)\|^2 \, dt \tag{6.7}$$

for the all $t_1 > t_0$ and all piecewise-continuous functions $w(t)$. In turn, a locally admissible controller (6.5) [respectively, (6.6)] constitutes a local solution of the $\mathcal{H}_\infty$-control problem if there exists a neighborhood U of the equilibrium such that inequality (6.7) is satisfied for all $t_1 > t_0$ and all piecewise-continuous functions

6.1 Synthesis of Time-Varying Systems over Continuous-Time Measurements

$w(t)$ for which the state trajectory of the corresponding closed-loop system, starting from the initial point $x(t_0) = 0$ [and $\xi(t_0) = 0$], remains in U for all $t \in [t_0, t_1]$.

The *nonsmooth $\mathcal{H}_\infty$-control problem* is to find a globally (locally) admissible state-feedback controller (6.5) in the full information case or a globally (locally) admissible output-feedback controller (6.6) under the available measurement (6.3) such that the $\mathcal{L}_2$-gain of the closed-loop system (6.1), driven by (6.5) or by (6.6), is (locally) less than γ. Solving the above problem under γ approaching the infimal achievable level γ^* in (6.7) yields a *(sub)optimal $\mathcal{H}_\infty$ controller* with the (sub)optimal disturbance attenuation level γ^* ($\gamma > \gamma^*$).

6.1.2 Dynamic Nonsmooth Hamilton–Jacobi–Isaacs Inequalities

Here are the hypotheses under which a global solution of the $\mathcal{H}_\infty$-control problem is derived:

H1. For some positive γ and some positive definite function $F(x)$, there exists a locally Lipschitz continuous positive definite decrescent, radially unbounded proximal solution $V(x, t)$ of the Hamilton–Jacobi–Isaacs inequality

$$\frac{\partial V}{\partial t} + \frac{\partial V}{\partial x} f(x,t) + \gamma^2 \alpha_1^T(x,t)\alpha_1(x,t) -$$
$$\alpha_2^T(x,t)\alpha_2(x,t) + h_1^T(x,t)h_1(x,t) + F(x) \leq 0, \qquad (6.8)$$

specified with

$$\alpha_1(x,t) = \frac{1}{2\gamma^2} g_1^T(x,t) \left(\frac{\partial V}{\partial x}\right)^T, \qquad (6.9)$$

$$\alpha_2(x,t) = -\frac{1}{2} g_2^T(x,t) \left(\frac{\partial V}{\partial x}\right)^T. \qquad (6.10)$$

H2. For some positive γ, some piecewise-continuous function $G(t)$, and some positive semidefinite function $Q(x, \xi)$ such that $Q(0, \xi)$ is positive definite, there exists a locally Lipschitz continuous positive semidefinite decrescent proximal solution $W(x, \xi, t)$ of the Hamilton–Jacobi–Isaacs inequality

$$\frac{\partial W}{\partial t} + \left(\frac{\partial W}{\partial x} \; \frac{\partial W}{\partial \xi}\right) f_e(x,\xi,t) + h_e^T(x,\xi,t)h_e(x,\xi,t)$$
$$+ \gamma^2 \phi^T(x,\xi,t)\phi(x,\xi,t) + Q(x,\xi) \leq 0, \qquad (6.11)$$

specified with

$$f_e^T(x,\xi,t) = ((f_e^1)^T, (f_e^2)^T),$$
$$f_e^1 = f(x,t) + g_1(x,t)\alpha_1(x,t) + g_2(x,t)\alpha_2(\xi,t),$$
$$f_e^2 = f(\xi,t) + g_1(\xi,t)\alpha_1(\xi,t) + g_2(\xi,t)\alpha_2(\xi,t)$$
$$+ G(t)(h_2(x,t) - h_2(\xi,t)), \qquad (6.12)$$

$$h_e(x,\xi,t) = \alpha_2(\xi,t) - \alpha_2(x,t), \qquad (6.13)$$

$$\phi(x,\xi,t) = \frac{1}{2\gamma^2} g_e^T(x,t) \begin{pmatrix} \left(\frac{\partial W}{\partial x}\right)^T \\ \left(\frac{\partial W}{\partial \xi}\right)^T \end{pmatrix}, \qquad (6.14)$$

$$g_e(x,t) = \begin{pmatrix} g_1(x,t) \\ G(t)k_{21}(x,t) \end{pmatrix}, \qquad (6.15)$$

and this proximal solution $W(x,\xi,t)$ is such that $W(0,\xi,t)$ is positive definite and radially unbounded.

6.1.3 Global State-Feedback Design

First, we'll address the full information case. Provided that Hypothesis H1 holds, a state-feedback solution of the $\mathcal{H}_\infty$-control problem is derived in terms of a proximal solution of the Hamilton–Jacobi–Isaacs inequality (6.8).

Theorem 21. *Consider system (6.1), (6.2) with Assumptions A1–A3. Let Hypothesis H1 be satisfied. Then the static state feedback*

$$u = \alpha_2(x,t) \qquad (6.16)$$

is globally admissible and the $\mathcal{L}_2$ gain of the closed-loop system (6.1), (6.16) is less than γ with respect to output (6.2).

Proof. For later use, let's introduce the multivalued function

$$H(x,w,u,t) = \frac{\partial V}{\partial t} + \frac{\partial V}{\partial x}[f(x,t) + g_1(x,t)w + g_2(x,t)u]$$
$$+ h_1^T(x,t)h_1(x,t) + u^T u - \gamma^2 w^T w, \qquad (6.17)$$

where $V(x,t)$ is an appropriate proximal solution of the Hamilton–Jacobi–Isaacs inequality (6.8) and $\left(\frac{\partial V}{\partial x}, \frac{\partial V}{\partial t}\right)^T \in \partial^P V(x,t)$. It is clear that (6.17) is quadratic in (w,u). Relations (6.9) and (6.10) result in

6.1 Synthesis of Time-Varying Systems over Continuous-Time Measurements

$$\left(\frac{\partial H}{\partial w}\right)_{(w,u)=(\alpha_1,\alpha_2)} = \frac{\partial V}{\partial x} g_1(x,t) - 2\gamma^2 \alpha_1^T = 0, \qquad (6.18)$$

$$\left(\frac{\partial H}{\partial u}\right)_{(w,u)=(\alpha_1,\alpha_2)} = \frac{\partial V}{\partial x} g_2(x,t) + 2\alpha_2^T = 0, \qquad (6.19)$$

and expanding the quadratic function $H(x, w, u, t)$ in Taylor series yields

$$H(x,w,u,t) = H(x,\alpha_1(x,t),\alpha_2(x,t),t) - \gamma^2 \|w - \alpha_1(x,t)\|^2$$
$$+ \|u - \alpha_2(x,t)\|^2, \qquad (6.20)$$

where $H(x,\alpha_1(x,t),\alpha_2(x,t),t) \leq -F(x)$, in accordance with Hypothesis H1. Hence,

$$H(x,w,u,t) \leq \|u - \alpha_2(x,t)\|^2 - \gamma^2 \|w - \alpha_1(x,t)\|^2 - F(x) \qquad (6.21)$$

and utilizing (6.17), (6.21), one obtains

$$\frac{\partial V}{\partial t} + \frac{\partial V}{\partial x}[f(x,t) + g_1(x,t)w + g_2(x,t)u] \leq \|u - \alpha_2(x,t)\|^2$$

$$- \gamma^2 \|w - \alpha_1(x,t)\|^2 - \|h_1(x,t)\|^2 - \|u\|^2 + \gamma^2 \|w\|^2 - F(x). \qquad (6.22)$$

If we set $w = 0$, it follows that

$$\frac{d}{dt} V(x(t),t) = DV(x(t),t;\dot{x}(t),1) \leq \frac{\partial V}{\partial t} + \frac{\partial V}{\partial x}\dot{x}(t)$$

$$= \frac{\partial V}{\partial t} + \frac{\partial V}{\partial x}[f(x(t),(t) + g_2(x,t)\alpha_2(x,t)] \leq -F(x(t)) \qquad (6.23)$$

on the trajectories of the disturbance-free closed-loop system (6.1), (6.16). Hypothesis H1 thus ensures that on the internal dynamics of the closed-loop system (6.1), (6.16), $V(x,t)$ is a strict descrescent radially unbounded Lyapunov function. By Lemma 9, these dynamics prove to be globally asymptotically stable.

To complete the proof, it remains to demonstrate that while being affected by an external disturbance, the closed-loop system (6.1), (6.16) has $\mathcal{L}_2$ gain less than γ with respect to the to-be-controlled output (6.2). For this purpose, let's integrate inequality (6.22) on the disturbed dynamics (6.1), (6.16) with $w \neq 0$ and $x(t_0) = 0$ to derive

$$\int_{t_0}^{t_1} \frac{d}{dt} V(x(t),t) \, dt \leq -\gamma^2 \int_{t_0}^{t_1} \|w(t) - \alpha_1(x(t),t)\|^2 \, dt$$

$$- \int_{t_0}^{t_1} F(x(t)) \, dt - \int_{t_0}^{t_1} \|\alpha_2(x(t),t)\|^2 \, dt$$

$$- \int_{t_0}^{t_1} \|h_1(x(t),t)\|^2 \, dt + \gamma^2 \int_{t_0}^{t_1} \|w(t)\|^2 \, dt. \qquad (6.24)$$

Since under the initial condition $x(t_0) = 0$, the left-hand side of the above inequality integrates to a nonnegative value $V(x(t_1), t_1) - V(x(t_0), t_0) = V(x(t_1), t_1)$, relation (6.24) ensures that the corresponding system output (6.2) satisfies (6.7) for all $t_1 > t_0$ and all piecewise-continuous functions $w(t)$. Theorem 21 is thus proved.

6.1.4 Global Output-Feedback Design

Hypotheses H1 and H2, coupled together, yield a solution of the $\mathcal{H}_\infty$-control problem under the only available measurement (6.3). A solution of the $\mathcal{H}_\infty$-control problem in this case is carried out in terms of proximal solutions of the Hamilton–Jacobi–Isaacs inequalities (6.8), (6.11). Coupled to the former inequality that has arisen in the state-feedback design, the latter inequality is now additionally involved to cover the output-injection synthesis to be made on-line for estimating the required state feedback over the available measurements.

Theorem 22. *Consider system (6.1)–(6.3) with Assumptions A1–A3. Let Hypotheses H1 and H2 be satisfied with a certain $\gamma > 0$. Then the causal dynamic output feedback*

$$\dot{\xi} = f(\xi, t) + g_1(\xi, t)\alpha_1(\xi, t) + g_2(\xi, t)\alpha_2(\xi, t) + G(t)[y(t) - h_2(\xi, t)],$$

$$u = \alpha_2(\xi, t) \tag{6.25}$$

is globally admissible under the available measurement (6.3) and the $\mathcal{L}_2$ gain of the closed-loop system (6.1), (6.25) is less than γ with respect to output (6.2).

Proof. To begin with, let's introduce the multivalued function

$$H_e(x, \xi, r, t) = \frac{\partial W}{\partial t} + \left(\frac{\partial W}{\partial x} \; \frac{\partial W}{\partial \xi} \right) [f_e(x, \xi, t) + g_e(x, t)r]$$

$$+ h_e^T(x, \xi, t) h_e(x, \xi, t) - \gamma^2 r^T r, \tag{6.26}$$

where $W(x, \xi, t)$ is an appropriate proximal solution of the Hamilton–Jacobi–Isaacs inequality (6.11) and $\left(\frac{\partial W}{\partial x}, \frac{\partial W}{\partial \xi}, \frac{\partial W}{\partial t} \right)^T \in \partial^P W(x, \xi, t)$. Clearly, (6.26) is quadratic in r. Then we have

$$\left(\frac{\partial H_e}{\partial r} \right)_{r=\phi(x,\xi,t)} = \left(\frac{\partial W}{\partial x} \; \frac{\partial W}{\partial \xi} \right) g_e(x, t) - 2\gamma^2 \phi^T(x, \xi, t) = 0,$$

and if we apply Hypothesis H2, it follows that $H_e(x, \xi, \phi(x, \xi, t), t) \leq -Q(x, \xi)$. Thereby, the quadratic function $H_e(x, \xi, r, t)$ is expanded in the Taylor series as follows:

$$H_e(x, \xi, r, t) = H_e(x, \xi, \phi(x, \xi, t), t) - \gamma^2 \|r - \phi(x, \xi, t)\|^2$$

$$\leq -\gamma^2 \|r - \phi(x, \xi, t)\|^2 - Q(x, \xi), \tag{6.27}$$

6.1 Synthesis of Time-Varying Systems over Continuous-Time Measurements

and, by virtue of (6.26), one establishes that

$$\frac{\partial W}{\partial t} + \left(\frac{\partial W}{\partial x} \ \frac{\partial W}{\partial \xi} \right) [f_e(x,\xi,t) + g_e(x,t)(w - \alpha_1(x,t))]$$

$$\leq -\gamma^2 \|w - \alpha_1(x,t) - \phi(x,\xi,t)\|^2$$

$$- \|\alpha_2(\xi,t) - \alpha_2(x,t)\|^2 + \gamma^2 \|w - \alpha_1(x,t)\|^2 - Q(x,\xi). \quad (6.28)$$

Let's now demonstrate that the function

$$U(x,\xi,t) = V(x,t) + W(x,\xi,t), \quad (6.29)$$

composed of the corresponding proximal solutions of the Hamilton–Jacobi–Isaacs inequalities (6.8) and (6.11), is a strict Lyapunov function of the closed-loop system (6.1), (6.25) with $w = 0$. Indeed, this function is locally Lipschitz continuous, positive definite, radially unbounded, and decrescent by construction. Moreover, applying Lemma 8 and employing relations (6.22) and (6.28), coupled together, yield

$$\frac{dU}{dt} \leq \frac{\partial V}{\partial t} + \frac{\partial V}{\partial x}[f(x,t) + g_1(x,t)w + g_2(x,t)u] + \frac{\partial W}{\partial t}$$

$$+ \left(\frac{\partial W}{\partial x} \ \frac{\partial W}{\partial \xi} \right) [f_e(x,\xi,t) + g_e(x,t)(w - \alpha_1(x,t))]$$

$$\leq \|u - \alpha_2(x,t)\|^2 - \gamma^2 \|\alpha_1(x,t)\|^2 - \|h_1(x,t)\|^2$$

$$- \gamma^2 \|\alpha_1(x,t) + \phi(x,\xi,t)\|^2 - \|\alpha_2(\xi,t) - \alpha_2(x,t)\|^2$$

$$+ \gamma^2 \|\alpha_1(x,t)\|^2 - \|u\|^2 - F(x) - Q(x,\xi)$$

$$\leq - \|h_1(x,t)\|^2 - \gamma^2 \|\alpha_1(x,t) + \phi(x,\xi,t)\|^2$$

$$- \|\alpha_2(\xi,t)\|^2 - F(x) - Q(x,\xi) \leq -F(x) - Q(x,\xi). \quad (6.30)$$

Thus, Lemma 9 is applicable to the internal dynamics of the closed-loop system (6.1), (6.25), and system (6.1), (6.25) is therefore internally globally asymptotically stable.

Finally, to establish (6.7), we differentiate (6.29) along the trajectories of the perturbed system (6.1), (6.25) with $w \neq 0$ and using (6.22), (6.28), we then obtain

$$\frac{dU}{dt} \leq - \|z(t)\|^2 + \gamma^2 \|w\|^2 - F(x) - Q(x,\xi) - \gamma^2 \|w - \alpha_1(x,t) - \phi(x,\xi,t)\|^2.$$

$$(6.31)$$

Clearly, (6.31) ensures that

$$\int_{t_0}^{t_1} (\gamma^2 \|w\|^2 - \|z(t)\|^2) dt \geq \int_{t_0}^{t_1} [F(x(t)) + Q(x(t), \xi(t))] dt +$$

$$U(x(t_1), \xi(t_1), t_1) - U(x(t_0), \xi(t_o), t_0) +$$

$$\gamma^2 \int_{t_0}^{t_1} \|w(t) - \alpha_1(x(t), t) - \phi(x(t), \xi(t), t)\|^2 dt > 0 \qquad (6.32)$$

for any trajectory of (6.1), (6.25) starting at $x(t_0) = 0, \xi(t_0) = 0$. Thus, the dynamic output-feedback compensator (6.25) is shown to be a solution of the $\mathcal{H}_\infty$-control problem in question. This completes the proof of Theorem 22.

Remark 3. Notably, Theorem 21 is straightforwardly extracted from Theorem 22 once the perfect state measurement $y = x$ is admitted in the latter theorem. Actually, Hypothesis H2 would then no longer be needed in Theorem 22 and the static state feedback (6.16) would stand for the dynamic output-feedback compensator (6.25).

It should be noted that both Theorems 21 and 22 ensure the existence of a stabilizing feedback controller under no stabilizability–detectability conditions, thereby removing extra work on the explicit verification of these conditions. Later on, these results are utilized when we locally solve the problems in question.

6.1.5 Local State-Space Solution

As in the local $\mathcal{L}_2$-gain analysis of Sect. 5.3, the function $f(x,t) = f_1(x,t) + f_2(x,t)$ is further decomposed into two components, and for technical reasons, nonsmooth nonlinearities are assumed to only be absorbed into the term $f_2(x,t)$, whereas the other terms, including $f_1(x,t)$, are assumed to be smooth enough. The following assumptions are thus additionally made.

A4. The function $f(x,t)$ admits representation $f(x,t) = f_1(x,t) + f_2(x,t)$, and in some neighborhood $U(0)$ of the origin $x = 0$, the functions $f_1(x,t)$, $g_1(x,t)$, $g_2(x,t)$, $h_1(x,t)$, $h_2(x,t)$, $k_{12}(x,t)$, and $k_{21}(x,t)$ are uniformly bounded in $t \in \mathbb{R}$ and twice continuously differentiable in $x \in U(0)$, whereas their first- and second-order state derivatives are piecewise-continuous and uniformly bounded in $t \in \mathbb{R}$ for all $x \in U(0)$.
A5. The vector $\zeta = 0$ is a proximal time-uniform supergradient of the components $f_{2i}(x,t)$, $i = 1, \ldots, n$, of the function $f_2(x,t)$ at $x = 0$ for almost all t; that is,

$$f_{2i}(x,t) \leq \sigma \|x\|^2 \qquad (6.33)$$

for some $\sigma > 0$, almost all $t \in \mathbb{R}$, and all $x \in U(0)$.

6.1 Synthesis of Time-Varying Systems over Continuous-Time Measurements

Under Assumptions A1–A5, coupled together, the corresponding Hamilton–Jacobi–Isaacs inequalities are subsequently linearized and a local solution of the nonsmooth $\mathcal{H}_\infty$-control problem is then obtained. The development involves the linear $\mathcal{H}_\infty$-control problem for the system

$$\dot{x} = A(t)x + B_1(t)w + B_2(t)u,$$

$$z = C_1(t)x + D_{12}(t)u,$$

$$y = C_2(t)x + D_{21}(t)w, \quad (6.34)$$

where

$$A(t) = \frac{\partial f_1}{\partial x}(0,t), \ B_1(t) = g_1(0,t), \ B_2(t) = g_2(0,t), \ C_1(t) = \frac{\partial h_1}{\partial x}(0,t),$$

$$C_2(t) = \frac{\partial h_2}{\partial x}(0,t), \ D_{12}(t) = k_{12}(0,t), \ D_{21}(t) = k_{21}(0,t). \quad (6.35)$$

As presented in Chap. 3, such a problem is well understood if the linear system (6.34) is stabilizable and detectable from u and y, respectively. Recall that under these assumptions, the following conditions are necessary and sufficient for a solution of the problem to exist:

C1. The equation

$$-\dot{P} = P(t)A(t) + A^T(t)P(t) + C_1^T(t)C_1(t)$$

$$+ P(t)\left[\frac{1}{\gamma^2}B_1B_1^T - B_2B_2^T\right](t)P(t) \quad (6.36)$$

possesses a uniformly bounded positive semidefinite symmetric solution $P(t)$ such that the system

$$\dot{x} = [A - (B_2B_2^T - \gamma^{-2}B_1B_1^T)P](t)x(t) \quad (6.37)$$

is exponentially stable.

C2. As is specified with $A_1(t) = A(t) + \frac{1}{\gamma^2}B_1(t)B_1^T(t)P(t)$, the equation

$$\dot{Z} = A_1(t)Z(t) + Z(t)A_1^T(t) + B_1(t)B_1^T(t) +$$

$$Z(t)\left[\frac{1}{\gamma^2}PB_2B_2^TP - C_2^TC_2\right](t)Z(t), \quad (6.38)$$

possesses a uniformly bounded positive semidefinite symmetric solution $Z(t)$, such that the system

$$\dot{x} = [A_1 - Z(C_2^TC_2 - \gamma^{-2}PB_2B_2^TP)](t)x(t) \quad (6.39)$$

is exponentially stable.

According to the time-varying strict bounded real Lemma 6, Conditions C1 and C2 ensure that there exists a positive constant ε_0 such that the system of the perturbed Riccati equations

$$-\dot{P}_\varepsilon = P_\varepsilon(t)A(t) + A^T(t)P_\varepsilon(t) + C_1^T(t)C_1(t)$$

$$+ P_\varepsilon(t)\left[\frac{1}{\gamma^2}B_1 B_1^T - B_2 B_2^T\right](t)P_\varepsilon(t) + \varepsilon I, \tag{6.40}$$

$$\dot{Z}_\varepsilon = A_\varepsilon(t)Z_\varepsilon(t) + Z_\varepsilon(t)A_\varepsilon^T(t) + B_1(t)B_1^T(t)$$

$$+ Z_\varepsilon(t)\left[\frac{1}{\gamma^2}P_\varepsilon B_2 B_2^T P_\varepsilon - C_2^T C_2\right](t)Z_\varepsilon(t) + \varepsilon I \tag{6.41}$$

has a unique uniformly bounded positive definite symmetric solution $(P_\varepsilon(t), Z_\varepsilon(t))$ for each $\varepsilon \in (0, \varepsilon_0)$, where $A_\varepsilon(t) = A(t) + \frac{1}{\gamma^2}B_1(t)B_1^T(t)P_\varepsilon(t)$. Equations (6.40) and (6.41) are now utilized to derive a local solution of the nonsmooth $\mathcal{H}_\infty$-control problem for system (6.1)–(6.3).

6.1.5.1 Full Information Case

The result stated next addresses the local $\mathcal{H}_\infty$ synthesis under perfect state measurements and relies on the perturbed Riccati equation (6.40) that arises in the $\mathcal{H}_\infty$ state-feedback design of the linearized system (6.34). The proposed synthesis is therefore of the same level of complexity as that in the linear case.

Theorem 23. *Consider system (6.1), (6.2) with Assumptions A1–A5. Let Condition C1 be satisfied for some $\gamma > 0$ and let $P_\varepsilon(t)$ be a uniformly bounded positive definite symmetric solution of (6.40) under some $\varepsilon > 0$. Then Hypotheses H1 holds locally around the equilibrium $x = 0$ with*

$$V(x,t) = x^T P_\varepsilon(t)x, \tag{6.42}$$

$$F(x) = \frac{\varepsilon}{2}\|x\|^2, \tag{6.43}$$

and the static state feedback

$$u = -g_2^T(x,t)P_\varepsilon(t)x \tag{6.44}$$

is a local solution of the $\mathcal{H}_\infty$-control problem with the disturbance attenuation level γ.

Proof. First, we demonstrate that in a neighborhood of the origin $x = 0$ and for almost all $t \in \mathbb{R}$, the function $V(x,t) = x^T P_\varepsilon(t)x$, being differentiable, positive definite, radially unbounded, and decrescent by construction, satisfies the Hamilton–Jacobi–Isaacs inequality (6.8) subject to (6.43). Indeed,

6.1 Synthesis of Time-Varying Systems over Continuous-Time Measurements

$$\frac{\partial V}{\partial t} = x^T \dot{P}_\varepsilon(t) x, \tag{6.45}$$

$$\frac{\partial V}{\partial x} f_1(x,t) = x^T [P_\varepsilon A + A^T P_\varepsilon](t) x + o_t(\|x\|^2), \tag{6.46}$$

$$\frac{\partial V}{\partial x} f_2(x,t) \leq \sigma^0 \|x\|^3, \tag{6.47}$$

$$\gamma^2 \alpha_1^T(x,t)\alpha_1(x,t) = \frac{1}{\gamma^2} x^T P_\varepsilon B_1(t) B_1^T(t) P_\varepsilon x + o_t(\|x\|^2), \tag{6.48}$$

$$-\alpha_2^T(x,t)\alpha_2(x,t) = -x^T P_\varepsilon B_2(t) B_2^T(t) P_\varepsilon x + o_t(\|x\|^2), \tag{6.49}$$

$$h_1^T(x,t) h_1(x,t) = x^T C_1^T(t) C_1(t) x + o_t(\|x\|^2), \tag{6.50}$$

where $\sigma^0 = 2\sigma \sup_{t \in \mathbb{R}} \|P_\varepsilon(t)\|$ and by virtue of Assumption A4, $\frac{o_t(\|x\|^2)}{\|x\|^2} \to 0$ uniformly in t as $\|x\|^2 \to 0$. Then due to (6.40), we have

$$\frac{\partial V}{\partial t} + \frac{\partial V}{\partial x}[f_1(x,t) + f_2(x,t)] + \gamma^2 \alpha_1^T(x,t)\alpha_1(x,t) -$$

$$\alpha_2^T(x,t)\alpha_2(x,t) + h_1^T(x,t)h_1(x,t) \leq$$

$$x^T \left\{ \dot{P}_\varepsilon + P_\varepsilon A + A^T P_\varepsilon + P_\varepsilon \left[\frac{1}{\gamma^2} B_1 B_1^T - B_2 B_2^T \right] P_\varepsilon + C_1^T C_1 \right\}(t) x +$$

$$\sigma^0 \|x\|^3 + o_t(\|x\|^2) \leq -\varepsilon \|x\|^2 + o_t(\|x\|^2) \leq -\frac{\varepsilon}{2}\|x\|^2 \tag{6.51}$$

for almost all $t \in \mathbb{R}$ and all $\|x\|$ sufficiently small. Hence, Hypothesis H1 holds locally with $V(x,t)$ and $F(x)$ defined by (6.42) and (6.43).

To complete the proof, it suffices to note that the line of reasoning used in the proof of Theorem 21 applies here as well. By following this line, we may conclude that the static state feedback (6.16), which, due to (6.42), (6.43), is given by (6.44), represents a local solution of the $\mathcal{H}_\infty$-control problem in question. Theorem 23 is thus proved.

6.1.5.2 Incomplete Information Case

Under partial state measurements, the local $\mathcal{H}_\infty$ synthesis is augmented with a dynamic compensator running in parallel. The compensator is derived by means of the perturbed Riccati equations (6.40), (6.41), appearing in the state-feedback design and, respectively, output-injection design of the linearized system (6.34). Thus, the resulting synthesis of the underlying nonsmooth system comes with the same level of complexity as that of the linearized system.

Theorem 24. *Consider system (6.1)–(6.3) with Assumptions A1–A5. Let Conditions C1 and C2 be satisfied with a certain $\gamma > 0$ and let $(P_\varepsilon(t), Z_\varepsilon(t))$ be a uniformly bounded positive definite symmetric solution of (6.40), (6.41) under some $\varepsilon > 0$. Then Hypotheses H1 and H2 hold locally around the equilibrium $(x, \xi) = (0, 0)$ with $V(x, t), F(x)$, given by (6.42), (6.43), and with*

$$W(x, \xi, t) = \gamma^2 (x - \xi)^T Z_\varepsilon^{-1}(t)(x - \xi), \tag{6.52}$$

$$G(t) = Z_\varepsilon(t) C_2^T(t), \tag{6.53}$$

$$Q(x, \xi) = \frac{\varepsilon}{2} \gamma^2 \min_{t \in \mathbb{R}} \|Z_\varepsilon^{-1}(t)\|^2 \|x - \xi\|^2. \tag{6.54}$$

Moreover, the causal dynamic output-feedback compensator

$$\dot{\xi} = f(\xi, t) + \left[\frac{1}{\gamma^2} g_1(\xi, t) g_1^T(\xi, t) - g_2(\xi, t) g_2^T(\xi, t) \right] P_\varepsilon(t) \xi$$

$$+ Z_\varepsilon(t) C_2^T(t)[y(t) - h_2(\xi, t)], \tag{6.55}$$

$$u = -g_2^T(\xi, t) P_\varepsilon(t) \xi \tag{6.56}$$

is a local solution of the $\mathcal{H}_\infty$-control problem with the disturbance attenuation level γ.

Proof. To begin with, let's note that the validity of Hypothesis H1 has been established by Theorem 23.

Next let's consider the function $W(x, \xi, t)$, given by (6.52). By construction, this function is smooth, positive semidefinite and decrescent and such that $W(0, \xi, t)$ is positive definite and radially unbounded. Moreover, in a neighborhood of the origin $(x, \xi) = (0, 0)$ and for almost all $t \in \mathbb{R}$, it satisfies the Hamilton–Jacobi–Isaacs inequality (6.11) subject to (6.53) and (6.54). Indeed,

$$\frac{\partial W}{\partial t} = \gamma^2 (x - \xi)^T \dot{Z}_\varepsilon^{-1}(t)(x - \xi), \tag{6.57}$$

$$\left(\frac{\partial W}{\partial x} \ \frac{\partial W}{\partial \xi} \right) f_e(x, \xi, t) \leq \gamma^2 (x - \xi)^T \{ Z_\varepsilon^{-1} A_\varepsilon + A_\varepsilon^T Z_\varepsilon^{-1} -$$

$$2 C_2^T C_2 \}(t)(x - \xi) + o_t(\|x - \xi\|^2), \tag{6.58}$$

$$h_e^T(x, \xi, t) h_e(x, \xi, t) = (x - \xi)^T [P_\varepsilon B_2 B_2^T P_\varepsilon](t)(x - \xi) + o_t(\|x - \xi\|^2), \tag{6.59}$$

6.1 Synthesis of Time-Varying Systems over Continuous-Time Measurements

$$\gamma^2 \phi^T(x, \xi, t) \phi(x, \xi, t) = \gamma^2 (x - \xi)^T \{ Z_\varepsilon^{-1} B_1 B_1^T Z_\varepsilon^{-1} +$$
$$C_2^T C_2 \}(t)(x - \xi) + o_t(\|x - \xi\|^2), \quad (6.60)$$

and taking into account the equation

$$-\dot{Z}_\varepsilon^{-1} = Z_\varepsilon^{-1}(t) A_\varepsilon(t) + A_\varepsilon^T(t) Z_\varepsilon^{-1}(t) + \left[\frac{1}{\gamma^2} P_\varepsilon B_2 B_2^T P_\varepsilon - \right.$$
$$\left. C_2^T C_2 \right](t) + Z_\varepsilon^{-1}(t) B_1(t) B_1^T(t) Z_\varepsilon^{-1}(t) + \varepsilon Z_\varepsilon^{-2}(t), \quad (6.61)$$

resulting from (6.41), one derives that

$$\frac{\partial W}{\partial t} + \left(\frac{\partial W}{\partial x} \; \frac{\partial W}{\partial \xi} \right) f_e(x, \xi, t) + h_e^T(x, \xi, t) h_e(x, \xi, t) + \gamma^2 \phi^T(x, \xi, t) \phi(x, \xi, t) \le$$

$$\gamma^2 (x - \xi)^T \left\{ \dot{Z}_\varepsilon^{-1} + Z_\varepsilon^{-1} A_\varepsilon + A_\varepsilon^T Z_\varepsilon^{-1} \left[\frac{1}{\gamma^2} P_\varepsilon B_2 B_2^T P_\varepsilon - C_2^T C_2 \right] + \right.$$

$$\left. Z_\varepsilon^{-1} B_1 B_1^T Z_\varepsilon^{-1} \right\} (t)(x - \xi) + o_t(\|x - \xi\|^2) =$$

$$-\varepsilon \gamma^2 (x - \xi)^T Z_\varepsilon^{-2}(t)(x - \xi) + o_t(\|x - \xi\|^2) \le -\frac{\varepsilon}{2} \gamma^2 \min_{t \in \mathbb{R}} \|Z_\varepsilon^{-1}(t)\|^2 \|x - \xi\|^2$$

for almost all $t \in \mathbb{R}$ and all $\|x\|, \|x - \xi\|$ sufficiently small. Hence, Hypothesis H2 holds locally with $W(x, \xi, t)$, $G(t)$, and $Q(x, \xi)$ defined by (6.52), (6.53), and (6.54), respectively.

As both Hypotheses H1 and H2 hold locally, following the line of reasoning used in the proof of Theorem 22, one may conclude that the output feedback (6.25), which, due to (6.42)–(6.54), is given by (6.55), (6.56), represents a local solution of the $\mathcal{H}_\infty$-control problem in question. Theorem 24 is proved.

6.1.5.3 $\mathcal{H}_\infty$-Design Procedure

The resulting local $\mathcal{H}_\infty$-design procedure consists of several steps. While being presented here for the case of incomplete state measurements, it is readily specified for the full information case by confining it to the design of the static controller (6.44) (cf. Remark 3). The overall procedure is as follows.

- First, γ large enough is selected such that Conditions $C1$ and $C2$ are satisfied.
- Then if we iterate on γ, the infimal achievable level γ^* is approached over all γ for which Conditions $C1$ and $C2$ are in force. By Theorem 24, (sub)optimal $\mathcal{L}_2$ gain of the underlying system is thus simultaneously achieved.

- After that a uniformly bounded, positive definite symmetric solution of the system of the perturbed differential Riccati equations (DREs) (6.40), (6.41) with $\gamma = \gamma^*$ and $\varepsilon > 0$ small enough is constructed.
- Finally, controller (6.55), (6.56), which corresponds to γ and ε, thus selected, completes the locally (sub-) optimal $\mathcal{H}_\infty$-control synthesis of the nonsmooth time-varying system (6.1)–(6.3).

According to the above procedure, one should use the iteration on γ to compute the $\mathcal{H}_\infty$ suboptimal solution as γ approaches the infimal achievable level γ^* in (6.7). By Theorem 24, γ^* may indeed be found as the infimum over all γ that were admissible at the first step. Once such an admissible γ^* is fixed at the second step, the next step does become feasible due to the time-varying strict bounded real Lemma 6. The final step of the procedure is then straightforward.

6.2 Local Output-Feedback Synthesis of Periodic and Autonomous Systems

In the periodic case or, particularly, in the autonomous case, where all the functions in (6.1)–(6.3) and (6.35) are time-periodic of a period $T > 0$ or time-independent, the interest is typically focused on the design of a time-periodic or time-invariant controller. Since Remark 3 applies here as well, the sequel deals with partial state measurements only, implying that the static state-feedback design can readily be extracted from the dynamic output-feedback design.

In the periodic case, the DREs (6.36), (6.38) come with the time-periodic matrices $A(t), B_i(t), C_i(t), i = 1, 2$, and Conditions $C1$ and $C2$ are modified to

$C1'$. The DRE (6.36) possesses a T-periodic positive semidefinite symmetric solution $P(t)$ such that system (6.37) is exponentially stable.

$C2'$. Specified with $A_1(t) = A(t) + \frac{1}{\gamma^2} B_1(t) B_1^T(t) P(t)$, the DRE (6.38) possesses a T-periodic positive semidefinite symmetric solution $Z(t)$ such that system (6.39) is exponentially stable.

According to the periodic strict bounded real Lemma 7, Conditions $C1'$ and $C2'$ ensure that there exists a positive constant ε_0 such that the system of the perturbed periodic Riccati equations (6.40) and (6.41) has a unique periodic positive definite symmetric solution $(P_\varepsilon(t), Z_\varepsilon(t))$ for each $\varepsilon \in (0, \varepsilon_0)$. Thus, in the periodic setting, Theorem 24 is represented as follows.

Theorem 25. *Let Conditions $C1'$ and $C2'$ be satisfied for system (6.1)–(6.3) with the time-periodic functions $f(x,t), k_{12}(x,t), k_{21}(x,t), g_i(x,t), h_i(x,t), i = 1, 2$, of a period $T > 0$ and let $(P_\varepsilon(t), Z_\varepsilon(t))$ be a T-periodic positive definite symmetric solution of (6.40), (6.41) under some $\varepsilon > 0$. Then the output feedback (6.55), (6.56) is a local T-periodic solution of the $\mathcal{H}_\infty$-control problem in the periodic case.*

Proof of Theorem 25 is nearly the same as that of Theorem 24 and is therefore omitted.

6.2 Local Output-Feedback Synthesis of Periodic and Autonomous Systems

In the autonomous case, the DREs (6.36), (6.38) degenerate to the algebraic Riccati equations (AREs) by setting $\dot{P} = 0, \dot{Z} = 0$ and Conditions $C1$ and $C2$ are simplified to

$C1''$. The equation

$$PA + A^T P + C_1^T C_1 + P \left[\frac{1}{\gamma^2} B_1 B_1^T - B_2 B_2^T \right] P = 0 \quad (6.62)$$

possesses a positive semidefinite symmetric solution P such that the matrix $A - (B_2 B_2^T - \gamma^{-2} B_1 B_1^T) P$ has all eigenvalues with negative real part.

$C2''$. Specified with $A_1 = A + \frac{1}{\gamma^2} B_1 B_1^T P$, the equation

$$A_1 Z + Z A_1^T + B_1 B_1^T + Z \left[\frac{1}{\gamma^2} P B_2 B_2^T P - C_2^T C_2 \right] Z = 0, \quad (6.63)$$

possesses a positive semidefinite symmetric solution Z such that the matrix $A_1 - Z(C_2^T C_2 - \gamma^{-2} P B_2 B_2^T P)$ has all eigenvalues with negative real part.

Conditions $C1''$ and $C2''$ are known from [38] to be necessary and sufficient for a solution of the linear $\mathcal{H}_\infty$-control problem for the time-invariant version of system (6.34) to exist. According to the strict bounded real lemma (see, e.g., [5]), Conditions $C1''$ and $C2''$ ensure that there exists a positive constant ε_0 such that the system of the perturbed AREs

$$P_\varepsilon A + A^T P_\varepsilon + C_1^T C_1 + P_\varepsilon \left[\frac{1}{\gamma^2} B_1 B_1^T - B_2 B_2^T \right] P_\varepsilon + \varepsilon I = 0, \quad (6.64)$$

$$A_\varepsilon Z_\varepsilon + Z_\varepsilon A_\varepsilon^T + B_1 B_1^T + Z_\varepsilon \left[\frac{1}{\gamma^2} P_\varepsilon B_2 B_2^T P_\varepsilon - C_2^T C_2 \right] Z_\varepsilon + \varepsilon I = 0 \quad (6.65)$$

has a unique positive definite symmetric solution $(P_\varepsilon, Z_\varepsilon)$ for each $\varepsilon \in (0, \varepsilon_0)$, where $A_\varepsilon = A + \frac{1}{\gamma^2} B_1 B_1^T P_\varepsilon$. Based on this solution, one can construct a time-invariant nonsmooth $\mathcal{H}_\infty$-controller as follows.

Theorem 26. *Let Conditions $C1''$ and $C2''$ be satisfied for system (6.1)–(6.3), which is assumed to be time-invariant, and let $(P_\varepsilon, Z_\varepsilon)$ be a positive definite symmetric solution of (6.64), (6.65) under some $\varepsilon > 0$. Then the time-invariant output feedback*

$$\dot{\xi} = f(\xi) + \left[\frac{1}{\gamma^2} g_1(\xi) g_1^T(\xi) - g_2(\xi) g_2^T(\xi) \right] P_\varepsilon \xi + Z_\varepsilon C_2^T [y - h_2(\xi)], \quad (6.66)$$

$$u = -g_2^T(\xi) P_\varepsilon \xi \quad (6.67)$$

is a local solution of the $\mathcal{H}_\infty$-control problem in the autonomous case.

Proof of Theorem 26 is similar to that of Theorem 24 and is therefore omitted.

6.2.1 $\mathcal{H}_\infty$-Design Procedure in Periodic and Time-Invariant Settings

The resulting local $\mathcal{H}_\infty$-design procedures in the periodic and autonomous cases are nearly the same as that obtained in the general time-varying setting, with the only exception that periodic and, respectively, time-invariant solutions of the corresponding Riccati equations are now due.

- First, γ large enough is selected such that Conditions $C1'$–$C2'$/$C1''$–$C2''$ are satisfied.
- Then by iterating on γ, one can determine the (sub-)optimal value γ^* such that Conditions $C1'$–$C2'$/$C1''$–$C2''$ are satisfied for all $\gamma > \gamma^*$.
- After that, a periodic/time-invariant positive definite symmetric solution of the system of the perturbed periodic/algebraic Riccati equations (6.40)–(6.41)/(6.64)–(6.65) with $\gamma = \gamma^*$ and $\varepsilon > 0$ small enough is constructed.
- Finally, controller (6.55)–(6.56)/(6.64)–(6.65), which corresponds to γ and ε, thus selected, completes the locally (sub-)optimal $\mathcal{H}_\infty$-control synthesis of the nonlinear periodic/time-invariant system (6.1)–(6.3).

6.3 Local Output-Feedback Synthesis over Sampled-Data Measurements

Let's now assume that sampled-data measurements are available at time instants $\tau_0 < \tau_1 < \cdots$ only and system (6.1)–(6.3) is thus specified as follows:

$$\dot{x}(t) = f(x(t), t) + g_1(x(t), t)w(t) + g_2(x(t), t)u(t), \quad (6.68)$$

$$z(t) = h_1(x(t), t) + k_{12}(x(t), t)u(t), \quad (6.69)$$

$$y(\tau_j) = h_2(x(\tau_j), \tau_j) + k_{21}(x(\tau_j), \tau_j)w(\tau_j), \; j = 0, 1, \ldots. \quad (6.70)$$

The goal of this section is to design a dynamic sampled-data measurement-feedback controller, with internal state $\xi \in \mathbb{R}^q$, which internally uniformly asymptotically stabilizes system (6.68)–(6.70) around the origin $(x, \xi) = (0, 0)$ and which is such that the closed-loop system has $\mathcal{L}_2/l_2$ gain less than γ.

Given a real number $\gamma > 0$, it is said that the closed-loop system (6.68)–(6.70) has $\mathcal{L}_2/l_2$ gain less than γ if the response z, resulting from w for initial state $x(t_0) = 0, \xi(t_0) = 0$, satisfies

$$\int_{t_0}^{t_1} \|z(t)\|^2 \, dt < \gamma^2 \left[\int_{t_0}^{t_1} \|w(t)\|^2 \, dt + \sum_{\tau_j \in [t_0, t_1]} \|w(\tau_j)\|^2 \right] \quad (6.71)$$

6.3 Local Output-Feedback Synthesis over Sampled-Data Measurements

for all $t_1 > t_0$ and all continuous functions $w(t)$. The right-hand side in (6.71) should be viewed as a mixed $\mathcal{L}_2/l_2$ norm on the uncertain signals affecting the system and the sampled-data measurements.

The nonsmooth $\mathcal{H}_\infty$-control problem over sampled-data measurements is to find a globally admissible controller (6.6) such that $\mathcal{L}_2/l_2$-gain of the closed-loop system (6.68)–(6.70) is less than γ. Since the $\mathcal{H}_\infty$ norm translated to the continuous- and discrete-time domains is nothing more than the $\mathcal{L}_2$- and l_2-induced norms, respectively, the above stated problem is a natural generalization of the standard $\mathcal{H}_\infty$-control problem.

In turn, a local solution to the above problem is defined as follows. A locally admissible controller (6.6) is said to be a local solution of the nonsmooth $\mathcal{H}_\infty$-control problem if there exists a neighborhood U of the origin such that inequality (6.71) is satisfied for all $t_1 > t_0$ and all piecewise-continuous functions $w(t)$ for which the state trajectory of the closed-loop system (6.6), (6.68)–(6.70), starting from the initial point $(x(t_0), \xi(t_0)) = (0, 0)$, remains in U for all $t \subset [t_0, t_1]$.

6.3.1 Main Result

Under Assumptions A1–A5, which are still assumed to be in force, we derive a local solution to the problem in question. The solution to be proposed invokes the differential equations

$$\dot{Z}_\varepsilon = \tilde{A}_\varepsilon(t) Z_\varepsilon(t) + Z_\varepsilon(t) \tilde{A}_\varepsilon^T(t) + B_1(t) B_1^T(t)$$
$$+ \gamma^{-2} Z_\varepsilon(t) P_\varepsilon(t) B_2(t) B_2^T(t) P_\varepsilon(t) Z_\varepsilon(t) + \varepsilon I, \qquad (6.72)$$

$$\dot{x} = [\tilde{A} + \gamma^{-2} Z P B_2 B_2^T P](t) x(t), \qquad (6.73)$$

$$\dot{\xi} = f(\xi, t) + \left[\frac{1}{\gamma^2} g_1(\xi, t) g_1^T(\xi, t) - g_2(\xi, t) g_2^T(\xi, t) \right] P_\varepsilon(t) \xi, \qquad (6.74)$$

with jumps

$$Z_\varepsilon(\tau_j+) = Z_\varepsilon(\tau_j-)[I + C_2^T(\tau_j) C_2(\tau_j) Z_\varepsilon(\tau_j-)]^{-1}, \qquad (6.75)$$

$$x(\tau_j+) = x(\tau_j-) - Z(\tau_j-)[I + C_2^T(\tau_j) C_2(\tau_j) Z(\tau_j-)]^{-1} C_2^T(\tau_j) C_2(\tau_j) x(\tau_j-), \qquad (6.76)$$

$$\xi(\tau_j+) = \zeta_j(1), \quad j = 0, 1, \ldots, \qquad (6.77)$$

where $\zeta_j(t)$ satisfies

$$\dot{\zeta}_j(t) = Z_\varepsilon(\tau_j-)[I + C_2^T(\tau_j) C_2(\tau_j) Z_\varepsilon(\tau_j-) t]^{-1} C_2^T(\tau_j) \times$$

$$[y_j - h_2(\zeta_j(t), \tau_j)], \quad \zeta_j(0) = \xi(\tau_j-). \qquad (6.78)$$

These jumps describe instantaneous changes in the dynamics of the $\mathcal{H}_\infty$ controller at the sampling time instants τ_j, $j = 0, 1, \ldots$. It should be noted that the controller output values $\xi(\tau_j+)$ from the right at τ_j, $j = 0, 1, \ldots$, are determined according to (6.77) through the evolution of the auxiliary dynamic system (6.78). Particularly when dealing with a linear observation (6.70), where $h_2(x,t) = C_2(t)x$, system (6.78) is analytically solved by means of

$$\zeta_j(t) = Z_\varepsilon(\tau_j-)[I + C_2^T(\tau_j)C_2(\tau_j)Z_\varepsilon(\tau_j-)t]^{-1}[Z_\varepsilon^{-1}(\tau_j-)\xi(\tau_j-) + C_2^T(\tau_j)y(\tau_j)],$$

and (6.77) is represented in the explicit form

$$\xi(\tau_j+) = Z_\varepsilon(\tau_j-)[I + C_2^T(\tau_j)C_2(\tau_j)Z_\varepsilon(\tau_j-)]^{-1}[Z_\varepsilon^{-1}(\tau_j-)\xi(\tau_j-) + C_2^T(\tau_j)y(\tau_j)]$$
$$= \xi(\tau_j-) + Z_\varepsilon(\tau_j-)[I + C_2^T(\tau_j)C_2(\tau_j)Z_\varepsilon(\tau_j-)]^{-1}C_2^T(\tau_j)[y_j - C_2(\tau_j)\xi(\tau_j-)],$$

similar to that for the variable Z_ε [cf. (6.75)].

In order to state the main result of this section, we modify Condition $C2$ to the following:

S2 Specified with $\varepsilon = 0$ and with a uniformly bounded positive semidefinite symmetric solution $P_{\varepsilon=0}(t) = P(t)$ of (6.36), system (6.72), (6.75) possesses a uniformly bounded positive semidefinite symmetric solution $Z(t)$ such that (6.73), (6.76) is exponentially stable.

The result, given below, locally solves the $\mathcal{H}_\infty$-control problem via a sampled-data–measurement feedback.

Theorem 27. *Let Conditions C1 and S2 be satisfied. Then there exists $\varepsilon_0 > 0$ such that for each $\varepsilon \in (0, \varepsilon_0)$, system (6.40), (6.72), (6.75) has a unique continuous from the left, uniformly bounded positive definite symmetric solution $(P_\varepsilon(t), Z_\varepsilon(t))$, and a solution of the nonsmooth $\mathcal{H}_\infty$-control problem for system (6.68)–(6.70) is given by (6.56), (6.74), (6.77), and (6.78).*

The proof of Theorem 27 appears in Sect. 6.3.3 after presenting a dual $\mathcal{H}_\infty$-control problem over continuous measurements that forms a basis of the proposed synthesis over sampled-data measurements.

6.3.2 Time Substitution-Based Transformation into the $\mathcal{H}_\infty$-Control Synthesis via Continuous Measurements

In order to prove Theorem 27, it suffices to transform the problem in question, by means of a certain time substitution, into a $\mathcal{H}_\infty$-control problem over continuous-time measurements. To reproduce the desired time substitution, let's introduce the functions

6.3 Local Output-Feedback Synthesis over Sampled-Data Measurements

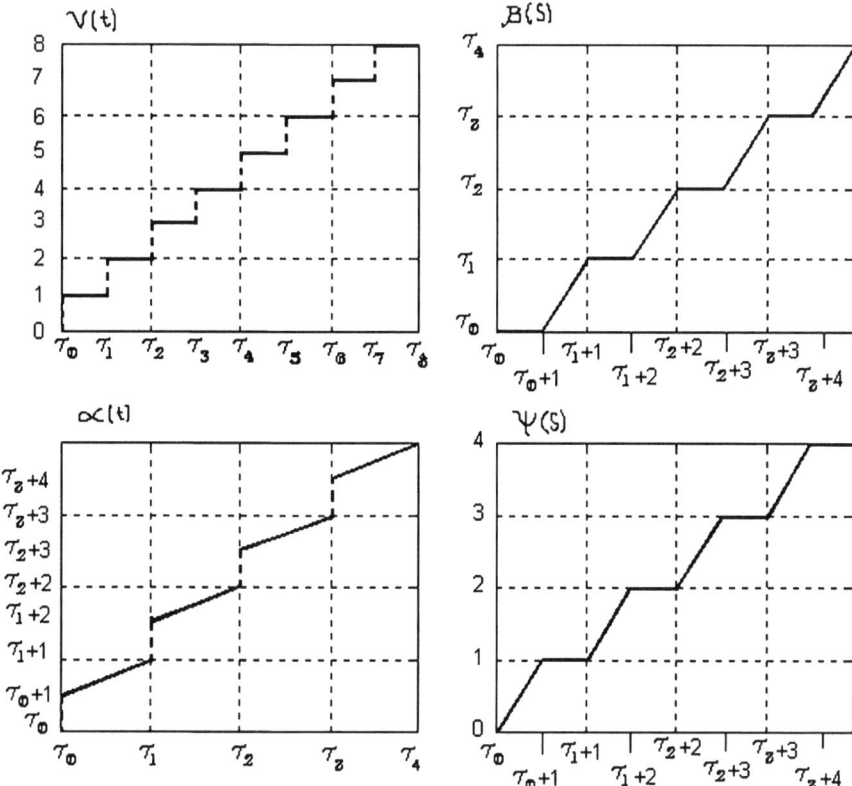

Fig. 6.1 Plots of the functions $v(t), \alpha(t), \beta(s)$, and $\psi(s)$

$$\chi(t) = \begin{cases} 0, & t \le 0 \\ 1, & t > 0 \end{cases}, \quad v(t) = \sum_{k=1}^{\infty} \chi(t - \tau_k), \tag{6.79}$$

$$\alpha(t) = t + v(t), \quad \beta(s) = \inf\{t : \alpha(t) > s\}, \tag{6.80}$$

$$\psi(s) = \begin{cases} s - t, & \text{if } \beta(s) = t \in D, \\ v(\beta(s)), & \text{otherwise,} \end{cases} \tag{6.81}$$

where $D = \{\tau_j\}_{j=0}^{\infty}$. For convenience of the reader, plots of these functions are depicted in Fig. 6.1.

It is clear that the functions $\beta(s)$ and $\psi(s)$ have the piecewise-continuous derivatives

$$\dot{\beta}(s) = \begin{cases} 0, & \text{if } \beta(s) \in D, \\ 1, & \text{otherwise,} \end{cases}, \quad \dot{\psi}(s) = \begin{cases} 1, & \text{if } \beta(s) \in D, \\ 0, & \text{otherwise,} \end{cases} \tag{6.82}$$

and

$$\beta(\alpha(t)) = t \text{ for all } t \in \mathbb{R}^1, \quad (6.83)$$

$$\alpha(\beta(s)) = s \text{ for all } s \in \mathbb{R} \text{ such that } \beta(s) \notin D,$$

$$\alpha(\beta(s)-) = \inf\{s' : \beta(s') = \beta(s)\} \text{ for all } s \in \mathbb{R} \text{ such that } \beta(s) \in D,$$

$$\alpha(\beta(s)+) = \alpha(\beta(s)-) + 1 \text{ for all } s \in \mathbb{R} \text{ such that } \beta(s) \in D. \quad (6.84)$$

Along with the $\mathcal{H}_\infty$-control problem for the nonlinear system (6.68), (6.69) over the sampled-data measurement (6.70), we shall consider the $\mathcal{H}_\infty$-control problem for the auxiliary system

$$\dot{\hat{x}}(s) = \hat{f}(\hat{x}(s), s) + \hat{g}_1(\hat{x}(s), s)\hat{w}(s) + \hat{g}_2(\hat{x}(s), s)\hat{u}(s),$$
$$\hat{z}(s) = \hat{h}_1(\hat{x}(s), s) + \hat{k}_{12}(\hat{x}(s), s)\hat{u}(s),$$
$$\hat{y}(s) = \hat{h}_2(\hat{x}(s), s) + \hat{k}_{21}(\hat{x}(s), s)\hat{w}(s), \quad (6.85)$$

with the continuous-time measurement $\hat{y}$ on the system. In the above equations, $\hat{x} \in \mathbb{R}^n$ is the state vector, $s \in \mathbb{R}$ is the time variable, $\hat{u} \in \mathbb{R}^m$ is the control input, $\hat{w} \in \mathbb{R}^r$ is the unknown disturbance, $\hat{z} \in \mathbb{R}^l$ is the unknown output to be controlled, and $\hat{y} \in \mathbb{R}^p$ is the only available continuous-time measurement on the system

$$\hat{f}(\hat{x}, s) = f(\hat{x}, \beta(s))\dot{\beta}(s), \quad \hat{g}_1(\hat{x}, s) = g_1(\hat{x}, \beta(s))\dot{\beta}(s),$$
$$\hat{g}_2(\hat{x}, s) = g_2(\hat{x}, \beta(s))\dot{\beta}(s), \quad \hat{h}_1(\hat{x}, s) = h_1(\hat{x}, \beta(s))\dot{\beta}(s),$$
$$\hat{h}_2(\hat{x}, s) = h_2(\hat{x}, \beta(s))\dot{\psi}(s), \quad \hat{k}_{12}(\hat{x}, s) = k_{12}(\hat{x}, \beta(s))\dot{\beta}(s),$$
$$\hat{k}_{21}(\hat{x}, s) = k_{21}(\hat{x}, \beta(s))\dot{\psi}(s). \quad (6.86)$$

As shown next, the auxiliary problem thus formed proves to be equivalent, in a certain sense, to the problem in question.

Theorem 28. *There exists a (local and/or exponentially stabilizing) solution of the sampled-data measurement $\mathcal{H}_\infty$-feedback control problem for the nonlinear system (6.68)–(6.70) if and only if there exists a (local and/or exponentially stabilizing) solution of the continuous-time measurement $\mathcal{H}_\infty$-feedback control problem for the nonlinear system (6.85) specified with (6.86). Moreover, if $u(t)$ is a (local and/or exponentially stabilizing) solution of the former problem, then $\hat{u}(s) = u(\beta(s))$ is a (local and/or exponentially stabilizing) solution of the latter problem provided that $\beta(s)$ is governed by (6.80). Conversely, if $\hat{u}(s)$ is a (local and/or exponentially*

6.3 Local Output-Feedback Synthesis over Sampled-Data Measurements

stabilizing) solution of the latter problem, then $u(t) = \hat{u}(\alpha(t))$ is a (local and/or exponentially stabilizing) solution of the former problem provided that $\alpha(t)$ is governed by (6.80).

Proof. Let $\hat{x}(s)$ be a trajectory of (6.85) driven by an admissible dynamic controller $\hat{u}(s)$ and subjected to an external disturbance $\hat{w}(s)$. Then it is straightforward to check that in accordance with (6.83) and (6.86), $\hat{x}(\alpha(t))$ is a trajectory of (6.68) driven by the admissible dynamic controller $u(t) = \hat{u}(\alpha(t))$ and subjected to the external disturbance $w(t) = \hat{w}(\alpha(t))$. By the same reasoning, if $x(t)$ is a trajectory of (6.68) enforced by $u(t), w(t)$, then $x(\beta(s))$ is a solution of (6.85) enforced by $\hat{u}(s) = u(\beta(s)), \hat{w}(s) = w(\beta(s))$. It follows that a (local and/or exponentially stabilizing) solution $u(t)$ of the sampled-data measurement $\mathcal{H}_\infty$-feedback control problem for (6.68)–(6.70) generates the (local and/or exponentially stabilizing) solution $\hat{u}(s) = u(\beta(s))$ of the continuous-time measurement $\mathcal{H}_\infty$-feedback control problem for (6.85), whereas a (local and/or exponentially stabilizing) solution $\hat{u}(s)$ of the latter problem generates the (local and/or exponentially stabilizing) solution $u(t) = \hat{u}(\alpha(t))$ of the former problem. Theorem 28 is thus proved.

We conclude this section by providing sufficient conditions for a local solution of the $\mathcal{H}_\infty$-control problem for the auxiliary system (6.85) to exist. These conditions are as follows:

CS1. The equation

$$-\dot{\hat{P}}(s) = \hat{P}(s)A(\beta(s))\dot{\beta}(s) + A^T(\beta(s))\hat{P}(s)\dot{\beta}(s) + C_1^T(\beta(s))C_1(\beta(s))\dot{\beta}(s)$$
$$+ \hat{P}(s)\left[\frac{1}{\gamma^2}B_1(\beta(s))B_1^T(\beta(s)) - B_2(\beta(s))B_2^T(\beta(s))\right]\hat{P}(s)\dot{\beta}(s) \quad (6.87)$$

possesses a uniformly bounded positive semidefinite symmetric solution $\hat{P}(s)$ such that the system

$$\dot{\hat{x}}(s) = \{A(\beta(s)) - [B_2(\beta(s))B_2^T(\beta(s))$$
$$- \gamma^{-2}B_1(\beta(s))B_1^T(\beta(s))]\hat{P}(s)\}\dot{\beta}(s)\hat{x}(s) \quad (6.88)$$

is exponentially stable.

CS2. Specified with $\hat{A}(s) = A(\beta(s)) + \frac{1}{\gamma^2}B_1(\beta(s))B_1^T(\beta(s))\hat{P}(s)$, the equation

$$\dot{\hat{Z}}(s) = \hat{A}(s)\hat{Z}(s)\dot{\beta}(s) + \hat{Z}(s)\hat{A}^T(t)\dot{\beta}(s) + B_1(\beta(s))B_1^T(\beta(s))\dot{\beta}(s) +$$
$$\hat{Z}(s)\left[\frac{1}{\gamma^2}\hat{P}(s)B_2(\beta(s))B_2^T(\beta(s))\hat{P}(s)\dot{\beta}(s) - C_2^T(\beta(s))C_2(\beta(s))\dot{\psi}(s)\right]\hat{Z}(s)$$

$$(6.89)$$

possesses a uniformly bounded positive semidefinite symmetric solution $\hat{Z}(s)$ such that the system

$$\dot{\hat{x}}(s) = \{\hat{A}(s)\dot{\beta}(s) - \hat{Z}(s)[C_2^T(\beta(s))C_2(\beta(s))\dot{\psi}(s)$$
$$-\gamma^{-2}\hat{P}(s)B_2(\beta(s))B_2^T(\beta(s))\hat{P}(s)\dot{\beta}(s)]\}\hat{x}(s) \qquad (6.90)$$

is exponentially stable.

The corresponding result is stated next in the form of a lemma.

Lemma 10. *Let Conditions CS1 and CS2 be satisfied. Then there exists $\varepsilon_o > 0$ such that system*

$$-\dot{\hat{P}}_\varepsilon(s) = \hat{P}_\varepsilon(s)A(\beta(s))\dot{\beta}(s) + A^T(\beta(s))\hat{P}_\varepsilon(s)\dot{\beta}(s) + C_1^T(\beta(s))C_1(\beta(s))\dot{\beta}(s)$$

$$+ \hat{P}_\varepsilon(s)\left[\frac{1}{\gamma^2}B_1B_1^T - B_2B_2^T\right](\beta(s))\hat{P}_\varepsilon(s)\dot{\beta}(s) + \varepsilon I, \qquad (6.91)$$

$$\dot{\hat{Z}}_\varepsilon(s) = \hat{A}(s)\hat{Z}_\varepsilon(s)\dot{\beta}(s) + \hat{Z}_\varepsilon(s)\hat{A}^T(s)\dot{\beta}(s) + B_1(\beta(s))B_1^T(\beta(s))\dot{\beta}(s)+$$

$$\hat{Z}_\varepsilon(s)\left[\frac{1}{\gamma^2}\hat{P}_\varepsilon(s)B_2(\beta(s))B_2^T(\beta(s))\hat{P}_\varepsilon(s)\dot{\beta}(s)-C_2^T(\beta(s))C_2(\beta(s))\dot{\psi}(s)\right]\hat{Z}_\varepsilon(s)+\varepsilon I \qquad (6.92)$$

has a unique uniformly bounded positive definite symmetric solution $(\hat{P}_\varepsilon(s), \hat{Z}_\varepsilon(s))$ for each $\varepsilon \in (0, \varepsilon_o)$, and a solution of the $\mathcal{H}_\infty$-control problem for the auxiliary system (6.85), (6.86) is given by

$$\dot{\hat{\xi}}(s) = \hat{f}(\hat{\xi}, s) + \left[\frac{1}{\gamma^2}\hat{g}_1(\hat{\xi}, s)\hat{g}_1^T(\hat{\xi}, s) - \hat{g}_2(\hat{\xi}, s)\hat{g}_2^T(\hat{\xi}, s)\right]\hat{P}_\varepsilon(s)\hat{\xi}$$

$$+ \hat{Z}_\varepsilon(s)C_2^T(\beta(s))\dot{\psi}(s)[\hat{y}(s) - \hat{h}_2(\hat{\xi}, s)], \qquad (6.93)$$

$$\hat{u} = -\hat{g}_2^T(\hat{\xi}, s)\hat{P}_\varepsilon(s)\hat{\xi}(s).$$

The proof of Lemma 10 follows the same line of reasoning as that of Theorems 22 and 24 and is therefore omitted. The above lemma is subsequently used in proving Theorem 27.

6.3.3 Proof of the Main Result

We first demonstrate that Conditions C1 and S2 are equivalent to Conditions CS1 and CS2, respectively.

6.3 Local Output-Feedback Synthesis over Sampled-Data Measurements

Indeed, if $\hat{P}(s)$, $\hat{Z}(s)$ are solutions of (6.87), (6.89), then it is straightforward to check that in continuity intervals of $\alpha(t)$, the functions $P(t) = \hat{P}(\alpha(t))$, $Z(t) = \hat{Z}(\alpha(t))$ satisfy (6.40), (6.72) subject to $\varepsilon = 0$. Moreover, for $s \in [\alpha(\tau_j -), \alpha(\tau_j +)]$ with $\tau_j \in D$ and $\alpha(\tau_j +) = \alpha(\tau_j -) + 1$, we have that $\dot{\beta}(s) = 0$, $\dot{\psi}(s) = 1$, and thus the following equations hold:

$$\dot{\hat{P}}(s) = 0,$$
$$\dot{\hat{Z}}(s) = -\hat{Z}(s)C_2^T(\tau_j)C_2(\tau_j)\hat{Z}(s), \qquad (6.94)$$

and the functions

$$\hat{P}(s) = \hat{P}(\alpha(\tau_j -)),$$
$$\hat{Z}(s) = \hat{Z}(\tau_j -)[I + C_2^T(\tau_j)C_2(\tau_j)Z(\tau_j -)s]^{-1} \qquad (6.95)$$

are solutions of (6.94). Thus, in spite of discontinuities in $\alpha(t)$, the function $\hat{P}(\alpha(t))$ is continuous for all t, thereby satisfying (6.40) for all t, whereas the jumps

$$\hat{Z}(\alpha(\tau_j +)) - \hat{Z}(\alpha(\tau_j -)) = \hat{Z}(\tau_j -)[I + C_2^T(\tau_j)C_2(\tau_j)Z(\tau_j -)]^{-1}$$
$$-\hat{Z}(\tau_j -), \quad j = 0, 1, \ldots,$$

of the function $\hat{Z}(\alpha(t))$ are the same as those of (6.72), (6.75), and therefore $\hat{Z}(\alpha(t))$ is a solution of the differential equation (6.72) with jumps (6.75).

Furthermore, if $(P(t), Z(t))$ is a solution of (6.40), (6.72), (6.75) subject to $\varepsilon = 0$, then by inspecting, one proves that $\hat{P}(s) = P(\beta(s))$, $\hat{Z}(s) = Z(\beta(s))$ satisfy (6.87), (6.89). To conclude the equivalence of Conditions CS1, CS2 and Conditions C1, S2, we still need to note that the same relations are also in force between solutions of (6.37), (6.73), (6.76) and (6.88), (6.90).

If we apply Theorem 28 and Lemma 10, it follows that Conditions C1, S2 are sufficient for a local solution of the sampled-data $\mathcal{H}_\infty$-feedback control problem for (6.68)–(6.70) to exist. Moreover, if these conditions are satisfied, Conditions CS1, CS2 are satisfied as well, and due to Lemma 10, there exists $\varepsilon_0 > 0$ such that system (6.91), (6.92) has a unique uniformly bounded positive definite symmetric solution $(\hat{P}_\varepsilon(s), \hat{Z}_\varepsilon(s))$ for each $\varepsilon \in (0, \varepsilon_o)$. By Lemma 10, these functions generate the unique uniformly bounded positive definite symmetric solution $(P_\varepsilon(t) = \hat{P}_\varepsilon(\alpha(t))$, $Z_\varepsilon(t) = \hat{Z}_\varepsilon(\alpha(t)))$ of (6.40), (6.72), (6.75) [the uniqueness is guaranteed by the invertibility of the time substitution in the above relations: $\hat{P}_\varepsilon(s) = P_\varepsilon(\beta(s))$, $\hat{Z}_\varepsilon(s) = Z_\varepsilon(\beta(s))$]. Thus, by Theorem 28, the continuous-time measurement-feedback solution (6.93), given by Lemma 10, generates the solution

$$u(t) = \hat{u}(\alpha(t)) = -\hat{g}_2^T(\hat{\xi}(\alpha(t)), \alpha(t))P_\varepsilon(\alpha(t))\hat{\xi}(\alpha(t))$$
$$= -g_2^T(\hat{\xi}(\alpha(t)), t)P_\varepsilon(t)\hat{\xi}(\alpha(t))$$

of the sampled-data $\mathcal{H}_\infty$-feedback control problem in question.

To complete the proof, let's demonstrate that the function $\xi(t) = \hat{\xi}(\alpha(t))$ satisfies the differential equation (6.74) with jumps (6.77), (6.78). In continuity intervals of $\alpha(t)$, governed by (6.80), we have $\dot{\psi}(s) = 0$, and by inspection, $\hat{\xi}(\alpha(t))$ is a solution of (6.74). If $s \in [\alpha(\tau_j-), \alpha(\tau_j+)]$, where $\tau_j \in D$ and $\alpha(\tau_j+) = \alpha(\tau_j-) + 1$, then the relations $\dot{\beta}(s) = 0$, $\dot{\psi}(s) = 1$, and (6.95) are in force and, consequently,

$$\dot{\hat{\xi}}(s) = Z_\varepsilon(\tau_j-)[I + C_2^T(\tau_j)C_2(\tau_j)Z_\varepsilon(\tau_j-)s]^{-1}C_2^T(\tau_j) \times$$

$$[y(\tau_j) - h_2(\hat{\xi}(s), \tau_j)], \quad \alpha(\tau_j-) \leq s \leq \alpha(\tau_j-) + 1.$$

Thus, the jumps $\hat{\xi}(\alpha(\tau_j+)) - \hat{\xi}(\alpha(\tau_j-))$, $j = 0, 1, \ldots$, of the function $\xi(t) = \hat{\xi}(\alpha(t))$ are the same as those defined by (6.77), (6.78), and therefore, $\hat{\xi}(\alpha(t))$ satisfies (6.74), (6.77), (6.78) for all t. Theorem 27 is proved.

6.3.4 $\mathcal{H}_\infty$-Design Procedure over Sampled-Data Measurements

The resulting local $\mathcal{H}_\infty$-design procedure over sampled-data measurements is similar to that developed over continuous-time measurements and is as follows.

- First, γ large enough is selected such that Conditions $C1$ and $S2$ are satisfied.
- Then, by iterating on γ's (sub-)optimal value, one can determine γ^* such that Conditions $C1$ and $S2$ are satisfied for all $\gamma > \gamma^*$.
- Next, a uniformly bounded positive definite symmetric solution of the system of the perturbed DREs (6.40), (6.72) with jumps (6.75) and with $\gamma = \gamma^*$ and $\varepsilon > 0$ small enough is constructed.
- Finally, controller (6.56), (6.74), (6.77), (6.78), which corresponds to γ and ε, thus selected, completes the locally (sub-)optimal $\mathcal{H}_\infty$-control synthesis of the nonlinear system (6.68), (6.69) over the sampled-data measurement (6.70).

Chapter 7
LMI-Based $\mathcal{H}_\infty$-Boundary Control of Nonsmooth Parabolic and Hyperbolic Systems

Many important plants, such as flexible manipulators and heat transfer processes, are governed by partial differential equations (PDEs) and are often described by models with a significant degree of uncertainty. The existing results [18, 35, 43, 65, 124] on robust control of distributed parameter systems (DPSs), operating under uncertainty conditions, extend the state-space or the frequency-domain $\mathcal{H}_\infty$ approach and are confined to the linear case. It is thus of interest to develop consistent methods that are capable of utilizing nonlinear distributed parameter models and of providing the desired system performance in spite of significant model uncertainties. The linear matrix inequality (LMI) approach, presented in Chap. 2, is definitely among such methods, and its extension to nonsmooth PDEs is expected to constitute an effective tool for robust control of DPSs.

The present chapter, which takes its inspiration from [51], develops exponential stability analysis and $\mathcal{L}_2$-gain analysis side by side for scalar uncertain DPSs governed by nonsmooth PDEs of the parabolic and hyperbolic types. The nonsmooth uncertainties are admitted to be time-, space-, and state-dependent, with a priori known upper and lower bounds. Sufficient exponential stability conditions with a given decay rate are derived in the form of LMIs for both systems. These conditions are then utilized to synthesize $\mathcal{H}_\infty$-static output-feedback boundary controllers of the systems in question. Numerical examples illustrate the efficacy of the method.

7.1 Boundary Stabilization of a Semilinear Parabolic System

Consider the parabolic equation

$$z_t(\xi,t) = \tfrac{\partial}{\partial \xi}[a(\xi) z_\xi(\xi,t)] + r_0(\xi,t,z(\xi,t))z(\xi,t) \\ + r_1(\xi,t,z(\xi,t))z(1,t), \ t \geq t_0, \ 0 \leq \xi \leq 1, \tag{7.1}$$

coupled to the mixed boundary condition

$$z(0,t) = 0, \quad z_\xi(1,t) = -kz(1,t), \ t \geq t_0, \quad (7.2)$$

where $t_0 \in \mathbb{R}$ is an initial time instant and $k \geq 0$ is a parameter. Functions $a(\xi)$ and $r_i(\xi,t,z)$, $i = 0,1$ are continuously differentiable and Lipschitz continuous, respectively, and these functions are unknown. These functions satisfy the inequalities

$$|r_i| \leq \beta_i, \ a \geq a_1 > 0, \quad i = 0,1 \quad (7.3)$$

for all $(\xi, t, z) \in [0,1] \times \mathbb{R}^2$ and for some constants $\beta_0 \geq 0, \beta_1 \geq 0, a_1 > 0$, known a priori. Hereinafter, the dependence on time t and spatial variable ξ is suppressed whenever possible and the functions a and r_i are written without arguments.

Subject to $r_1 \equiv 0$, Eq. (7.1) describes the nonlinear propagation of heat in a one-dimensional rod. A heat equation with the term $z(1,t)$ may describe the deviation from the steady state if the steady state depends on the boundary value (similar to that of [22]). Due to the presence of the boundary-value term $z(1,t)$ in the state equation (7.1) with $r_1 \neq 0$, the above model particularly captures significant features of thermal instability in solid propellant rockets [22].

Clearly, the boundary-value problem (7.1), (7.2) can be rewritten as the differential equation

$$\dot{x}(t) = Ax(t) + F(t, x(t)), \quad t \geq t_0, \quad (7.4)$$

in the Hilbert space $\mathcal{H} = L_2(0,1)$, where the infinitesimal operator $A = \frac{\partial [a(\xi) \frac{\partial}{\partial \xi}]}{\partial \xi}$ possesses the dense domain

$$\mathcal{D}(A) = \{x \in W^{2,2}(0,1) : x(0) = 0, x_\xi(1) = -kx(1)\}, \quad (7.5)$$

and the nonlinear term $F : \mathbb{R} \times W^{1,2}(0,1) \to L_2(0,1)$ is defined on potential solutions $x(\cdot, t)$ of (7.4) according to

$$F(t, x(\cdot)) = r_0(\xi, t, x(\xi, t))x(\xi, t)$$
$$+ r_1(\xi, t, x(\xi, t)) \int_0^1 x_\zeta(\zeta, t) d\zeta.$$

It is well known that the infinitesimal operator A generates an analytical exponentially stable semigroup $T(t)$, the induced norm $\|T(t)\|$ of which satisfies the inequality $\|T(t)\| \leq \kappa e^{-\delta t}$ everywhere with some constant $\kappa > 0$ and decay rate $\delta > 0$ (see, e.g., [35] for details). The domain $\mathcal{D}(A)$ of such an operator A forms another Hilbert space with the graph inner product $(x, y)_{\mathcal{D}(A)} = \langle Ax, Ay \rangle$, $x, y \in \mathcal{D}(A)$. The domain $\mathcal{D}(A)$ of A is thus continuously embedded into $\mathcal{H}$; namely, $\mathcal{D}(A) \subset \mathcal{H}$, $\mathcal{D}(A)$ is dense in $\mathcal{H}$ and the inequality $|x| \leq \omega |Ax|$ holds for all $x \in \mathcal{D}(A)$ and some constant $\omega > 0$.

7.1 Boundary Stabilization of a Semilinear Parabolic System

Apart from this, the square root $\sqrt{A}$ of the operator A is rigorously introduced on $\mathcal{D}(A)$ as a positive definite solution X of the algebraic operator equation $X^2 = A$. Extended by continuity, this operator is well posed on the domain

$$\mathcal{D}(\sqrt{A}) = \{x \in W^{1,2}(0,1) : x(0) = 0, x_\xi(1) = -kx(1)\}, \quad (7.6)$$

and continuously embedded into $\mathcal{H}$, whereas $\mathcal{D}(A)$ turns out to be continuously embedded into $\mathcal{D}(\sqrt{A})$. Thus, $\mathcal{D}(A) \subset \mathcal{D}(\sqrt{A}) \subset \mathcal{H}$ and the inequalities

$$|x| \leq \omega |\sqrt{A}x| \quad \text{for all } x \in \mathcal{D}(\sqrt{A}), \quad (7.7)$$

$$|\sqrt{A}x| \leq \omega |Ax| \quad \text{for all } x \in \mathcal{D}(A), \quad (7.8)$$

hold with a generic constant $\omega > 0$. All relevant background material on fractional operator degrees can be found, for instance, in [74].

Since the functions r_0 and r_1 are Lipschitz continuous in their arguments, the Lipschitz condition

$$\begin{aligned} &\|F(t_1, x_1) - F(t_2, x_2)\|_{L_2} \\ &\leq L[|t_1 - t_2| + \|\sqrt{A}(x_1 - x_2)\|_{L_2}] \end{aligned} \quad (7.9)$$

on the nonlinear term F with some positive constant L is derived locally in $(t_i, x_i) \in \mathbb{R} \times \mathcal{D}(\sqrt{A})$, $i = 1, 2$ if we use Wirtinger's inequalities (2.1), (2.2). Thus, [62, Theorem 3.3.3] proves to be applicable to (7.4), and by applying this theorem, we establish the local existence of a unique strong solution of (7.4), initialized with $x(t_0) \in \mathcal{D}(\sqrt{A})$ The latter implies the local existence of the strong solution to the boundary-value problem (7.1), (7.2) for an arbitrary initial condition

$$z(\xi, t_0) = \phi(\xi) \in \mathcal{D}(\sqrt{A}). \quad (7.10)$$

For convenience of the reader, recall that a continuous function $x(t)$, defined on some $[t_0, t_1)$, is a strong solution of the Hilbert space-valued differential equation (7.4), initialized with $x(t_0) \in \mathcal{D}(A)$, iff $\lim_{t \downarrow t_0} \|x(t) - x(t_0)\|_{\mathcal{H}} = 0$, and $x(t)$ is continuously differentiable and satisfies (7.4) for $t \in (t_0, t_1)$. Throughout, only strong solutions are under study.

7.1.1 Exponential Stability

It is well known (see, e.g., [22]) that the linear system (7.1), (7.2), with $k = 0$ and with constant coefficients $r_0 = 0, a = 1$, and $r_1 > 2$, is unstable. For the purpose of carrying out exponential stability conditions for the uncertain nonlinear system (7.1), (7.2) with $k \geq 0$, consider the following Lyapunov–Krasovskii functional:

$$V(z(\cdot, t)) = \int_0^1 z^2(\xi, t) d\xi. \quad (7.11)$$

The aim is to find conditions guaranteeing that along the solutions $z(\xi, t)$ of (7.1), (7.2), the inequality

$$\frac{d}{dt}V(z(\cdot,t)) + 2\delta V(z(\cdot,t)) \leq 0 \qquad (7.12)$$

holds. By the comparison principle argument [71], it would then follow that

$$\int_0^1 z^2(\xi,t)d\xi = V(z(\cdot,t)) \leq V(z(\cdot,t_0))e^{-2\delta(t-t_0)}$$
$$= e^{-2\delta(t-t_0)} \int_0^1 \phi^2(\xi)d\xi.$$

With this in mind, we see that the solution of (7.1), (7.2), (7.10) would be uniformly bounded in $L_2(0, 1)$ on its domain, and due to (7.3), it would satisfy the following inequality

$$\|F(t,x)\|_{L_2} \leq K[1 + \|\sqrt{A}x\|_{L_2}] \qquad (7.13)$$

for all $t \geq t_0$ and some positive constant K. Then according to [62, p. 58, Exercise 1], such a solution of the boundary-value problem (7.1), (7.2) would be globally continuable to the right. Hence, this solution would satisfy the inequality

$$\|z(\cdot,t)\|_{L_2} \leq e^{-\delta(t-t_0)}\|\phi(\cdot)\|_{L_2} \quad \text{for all} \quad t \geq t_0, \qquad (7.14)$$

thereby ensuring that the parabolic process (7.1), (7.2) is exponentially stable in $L_2(0, 1)$ with the decay rate δ.

Differentiating V along (7.1), integrating by parts, and taking into account (7.2) yield

$$\begin{aligned}
\frac{d}{dt}V + 2\delta V &= 2\int_0^1 z(\xi,t)z_t(\xi,t)d\xi + 2\delta\int_0^1 z^2(\xi,t)d\xi \\
&= 2\int_0^1 z(\xi,t)\left[\frac{\partial}{\partial \xi}[az_\xi(\xi,t)] + r_0 z(\xi,t) + r_1 z(1,t)\right]d\xi \\
&\quad + 2\delta\int_0^1 z^2(\xi,t)d\xi = -2kaz^2(1,t) - 2\int_0^1 az_\xi^2(\xi,t)d\xi \\
&\quad + 2\int_0^1 (\delta + r_0)z^2(\xi,t)d\xi + 2r_1\int_0^1 z(\xi,t)z(1,t)d\xi \\
&\leq -2ka_1 z^2(1,t) - 2\int_0^1 a_1 z_\xi^2(\xi,t)d\xi \\
&\quad + 2(\delta + \beta_0)\int_0^1 z^2(\xi,t)d\xi + 2r_1\int_0^1 z(\xi,t)z(1,t)d\xi.
\end{aligned} \qquad (7.15)$$

Since by Wirtinger's inequality,

$$-2a_1\int_0^1 z_\xi^2 d\xi \leq -4a_1\int_0^1 z^2(\xi,t)d\xi,$$

it follows that

$$\begin{aligned}
\frac{d}{dt}V + 2\delta V &\leq \int_0^1 [z(\xi,t)\ z(1,t)]\Psi[z(\xi,t)\ z(1,t)]^T d\xi \\
&\leq 0,
\end{aligned} \qquad (7.16)$$

7.1 Boundary Stabilization of a Semilinear Parabolic System

provided that the following LMI

$$\Psi \triangleq \begin{bmatrix} -4a_1 + 2(\delta + \beta_0) & r_1 \\ r_1 & -2ka_1 \end{bmatrix} \leq 0 \tag{7.17}$$

is feasible. Taking into account that LMI (7.17) is affine in r_1 and $r_1 \in [-\beta_1, \beta_1]$, we find that the latter LMI becomes feasible if the following LMI

$$\begin{bmatrix} -4a_1 + 2(\delta + \beta_0) & \beta_1 \\ \beta_1 & -2ka_1 \end{bmatrix} \leq 0 \tag{7.18}$$

is feasible.

It is worth noticing that the condition $\beta_0 < 2a_1$ is necessary for the feasibility of (7.18). Provided that $\beta_1 > 0$, the boundary-value problem (7.1), (7.2) is exponentially stable with the decay rate $0 < \delta < 2a_1 - \beta_0$ for large enough $k > 0$ that can be found from the inequality

$$-4a_1 + 2(\delta + \beta_0) + \frac{\beta_1^2}{2ka_1} \leq 0. \tag{7.19}$$

Once $\beta_1 = 0$, the boundary-value problem (7.1), (7.2) becomes exponentially stable for all $k \geq 0$ with the decay rate $\delta = 2a_1 - \beta_0$.

Summarizing, we may conclude the following result.

Theorem 29. *Consider the boundary-value problem (7.1), (7.2) under the initial condition (7.10) with the assumptions above and with $\beta_0 < 2a_1$. Given $\delta \in (0, 2a_1 - \beta_0]$, let there exist k such that LMI (7.18) is feasible. Then a unique strong solution of (7.1), (7.2), (7.10) is globally continuable to the right and satisfies (7.14).*

The above result forms a basis of the $\mathcal{H}_\infty$-boundary synthesis to be presented next.

7.1.2 $\mathcal{H}_\infty$-Boundary Control

Along with the homogeneous parabolic process (7.1), let's consider its perturbed version

$$z_t(\xi, t) = \frac{\partial}{\partial \xi}[a z_\xi(\xi, t)] + r_0 z(\xi, t) + r_1 z(1, t)$$

$$+ bw(\xi, t), \; t \geq t_0, \; 0 \leq \xi \leq 1, \tag{7.20}$$

where $w(\xi, t) \in L_2(0, \infty; L_2(0, 1))$ is an external disturbance; $b = b(\xi, t, z)$ is a Lipschitz continuous function, which is assumed to be uniformly bounded, namely, $|b(\xi, t, z)| \leq b_1$ for all $(\xi, t, z) \in [0, 1] \times \mathbb{R}^2$ and some $b_1 > 0$.

The influence of the admissible external disturbance $w(\xi, t) \in L_2(0, \infty; L_2(0, 1))$ on the controlled output

$$\bar{z}(\xi, t) = [\alpha(\xi, t, z(\xi, t))z(\xi, t) \quad d(t, z(1, t))u(t)]^T, \tag{7.21}$$

is to be attenuated through the boundary actuation

$$z(0, t) = 0, \quad z_\xi(1, t) = u(t), \ t \geq t_0, \tag{7.22}$$

at the right end $\xi = 1$ while the parabolic process is being internally stabilized. Hereinafter, $L_2(0, \infty; L_2(0, 1))$ is the Hilbert space of square-integrable functions on $(0, \infty)$ with values in $L_2(0, 1)$, whereas $u(t)$ is the control input, d and α are continuous functions, which are uniformly bounded,

$$|\alpha(\xi, t, z)| \leq \alpha_1, \ |d(t, z)| \leq d_1, \tag{7.23}$$

for all $(\xi, t, z) \in [0, 1] \times \mathbb{R}^2$ and some constants $\alpha_1 \geq 0$ and $d_1 \geq 0$. Collocated sensing $y(t) = z(1, t)$ at the boundary $\xi = 1$ is the only available information on the process.

The following $\mathcal{H}_\infty$-control problem is thus under study. Given $\gamma > 0$, we are required to find a linear static output feedback

$$u(t) = -kz(1, t) \tag{7.24}$$

that exponentially stabilizes the unperturbed process (7.2), (7.20) and leads to a negative performance index

$$J = \int_{t_0}^{\infty} \int_0^1 [\bar{z}^T(\xi, t)\bar{z}(\xi, t) - \gamma^2 w^2(\xi, t)] d\xi dt < 0 \tag{7.25}$$

for all the solutions of (7.20), (7.22), initialized with the zero data $z(\xi, t_0) = 0$, and for all the admissible external disturbances $0 \neq w(\xi, t) \in L_2(t_0, \infty; L_2(0, 1))$, under which these solutions are globally continuable to the right. Note that if $u(t)$ is stabilizing and w is Lipschitz continuous in (ξ, t), then by arguments of the previous section, the strong solutions of (7.20), (7.22) exist and they are continuable for $t \geq t_0$.

In order to solve the problem, one proposal is to carry out conditions that guarantee the following:

$$W(t) \triangleq p\frac{d}{dt}V + \int_0^1 [\bar{z}^T(\xi, t)\bar{z}(\xi, t) - \gamma^2 w^2(\xi, t)] d\xi < 0, \tag{7.26}$$

where $p > 0$, V is given by (7.11), and the temporal derivative is computed along the trajectories of the closed-loop system (7.2), (7.20). Then, integrating (2.20) in t from t_0 to ∞ and taking into account that $V \geq 0$ and $V(0) = 0$ will yield (7.25).

7.1 Boundary Stabilization of a Semilinear Parabolic System

It is worth noticing that

$$\int_0^1 \bar{z}^T(\xi,t)\bar{z}(\xi,t)d\xi \leq \int_0^1 \alpha_1^2 z^2(\xi,t)d\xi + d_1^2 k^2 z^2(1,t).$$

Then using (7.16), (7.17) and setting $\zeta = [z(\xi,t) \; z(1,t) \; w(\xi,t)]^T$ yield

$$W \leq \int_0^1 \zeta^T \Psi_\gamma \zeta d\xi < 0$$

provided that

$$\Psi_\gamma \triangleq \begin{bmatrix} -4a_1 p + 2\beta_0 p + \alpha_1^2 & r_1 p & bp \\ * & -2kap + d_1^2 k^2 & 0 \\ * & * & -\gamma^2 \end{bmatrix} < 0 \quad (7.27)$$

is feasible. Throughout, the symbol $*$ stands for the corresponding symmetric term of the matrix in question. By Schur complements, the latter inequality holds if

$$\begin{bmatrix} -4a_1 p + 2\beta_0 p + \alpha_1^2 & r_1 p & bp & 0 \\ * & -2ka_1 p & 0 & d_1 k \\ * & * & -\gamma^2 & 0 \\ * & * & * & -1 \end{bmatrix} < 0. \quad (7.28)$$

Multiplying (7.28) by $diag\{p^{-1}, p^{-1}, 1, 1\}$ from the right and from the left and then employing the Schur complements' formula, let's denote $q = p^{-1}$ and $g = p^{-1}k$ to arrive at

$$\begin{bmatrix} -4a_1 q + 2\beta_0 q & r_1 & b & 0 & q\alpha_1^2 \\ * & -2ga_1 & 0 & d_1 g & 0 \\ * & * & -\gamma^2 & 0 & 0 \\ * & * & * & -1 & 0 \\ * & * & * & * & -\alpha_1^2 \end{bmatrix} < 0. \quad (7.29)$$

The LMI (7.29) is affine in r_1 and b and is therefore feasible for all $r_1 \in [-\beta_1, \beta_1]$, $b \in [-b_1, b_1]$ whenever it is feasible for $r_1 = \pm\beta_1$ and $b = \pm b_1$, thereby yielding four LMIs. It is easy to see that these four LMIs are equivalent to the following LMI:

$$\begin{bmatrix} -4a_1 q + 2\beta_0 q & \beta_1 & b_1 & 0 & q\alpha_1^2 \\ * & -2ga_1 & 0 & d_1 g & 0 \\ * & * & -\gamma^2 & 0 & 0 \\ * & * & * & -1 & 0 \\ * & * & * & * & -\alpha_1^2 \end{bmatrix} < 0. \quad (7.30)$$

The following result is thus proved.

Theorem 30. *Consider the perturbed input–output system (7.20)–(7.22) with the assumptions above and with $\beta_0 < 2a_1$. Given $\gamma > 0$, let there exist $q > 0$ and g such that the LMI (7.30) is satisfied. Then the static output feedback (7.24) with $k = q^{-1}g$ internally exponentially stabilizes the boundary-value problem (7.20), (7.22) and attenuates the admissible disturbances $w(\xi,t) \in L_2(0,\infty; L_2(0,1))$ in the sense of (7.25).*

Analyzing the lines of reasoning of Theorems 29 and 30, one can note that these results remain in force even if the coefficient $a(\xi)$ depends not only on the spatial variable ξ but additionally depends on both the temporal variable t and the state variable z. The present investigation is, however, confined to the autonomous state-independent coefficient $a(\xi)$ to avoid the extra work of verifying the existence and uniqueness of strong solutions that are far from being routine manipulations in the case of a nonautonomous state-dependent coefficient $a(\xi, t, z)$.

Example 5. To illustrate the above synthesis, let's specify (7.20)–(7.23) with

$$a_1 = 1, \; b_1 = 1, \; \beta_0 = 1, \; \beta_1 = 3, \; d_1 = 0.1, \; \alpha_1 = 1.$$

Then the open-loop system with $u = 0$ and $w = 0$ is unstable [22] because $\beta_1 > 2a_1$. Since $\beta_0 < 2a_1$ and $\beta_1 > 0$, Theorem 29 can be applied to establish that the disturbance-free system is exponentially stabilized by the static output feedback (7.24) with a sufficiently large $k > 0$. By using MATLAB®'s LMI toolbox, we can verify the feasibility of the LMI (7.30) and obtain the static output feedback (7.24) with $k = 10.1744$, in accordance with Theorem 30, to additionally guarantee the disturbance attenuation level $\gamma = 3$. Substituting the resulting k into (7.19) yields the decay rate $\delta = 0.7789$ of the closed-loop system. A lower disturbance attenuation level $\gamma = 1.1$ is achieved with a higher gain $k = 106.01$. The decay rate, resulting from the latter gain, is found to be $\delta = 0.9788$.

7.2 Boundary Stabilization of a Semilinear Hyperbolic Equation

Consider the hyperbolic equation

$$\begin{aligned}z_{tt}(\xi,t) &= \tfrac{\partial}{\partial \xi}[az_\xi(\xi,t)] + r_0 z_t(\xi,t) \\ &+ r_1 z_\xi(\xi,t), \\ t &\geq t_0, \; 0 \leq \xi \leq 1,\end{aligned} \quad (7.31)$$

coupled with the mixed boundary condition

$$z(0,t) = 0, \; z_\xi(1,t) = -k z_t(1,t), \; t \geq 0, \quad (7.32)$$

where $k > 0$ is a parameter, $a = a(\xi)$, $r_0 = r_0(\xi, t, z, z_t)$, and $r_1 = r_1(\xi)$ are Lipschitz continuous functions. Subject to $r_1 \equiv 0$, Eq. (7.31) describes nonlinear

7.2 Boundary Stabilization of a Semilinear Hyperbolic Equation

oscillations of a string, whereas its general form is of academic interest. As in the parabolic equation (7.1), the functions r_i, $i = 0, 1$ are admitted to be unknown subject to inequalities (7.3), which hold for all $(\xi, t, z, z_t) \in [0, 1] \times \mathbb{R}^3$ with a priori known constants $\beta_i \geq 0, i = 0, 1$. Function a satisfies the bound

$$0 < a_1 \leq a(\xi) \leq a_2, \quad \forall \xi \in [0, 1], \tag{7.33}$$

with a priori known constants a_1, a_2.

To facilitate exposition, we have ignored the restoring stiffness of the string by implicitly assuming that the corresponding term $r(\xi, t, z, z_t)z(\xi, t)$ is negligible. Since the above simplified model captures all the essential features of the general treatment, the extension to a hyperbolic model with a nontrivial stiffness is indeed possible.

The boundary-value problem (7.31), (7.32) can be represented as the differential equation

$$\dot{x}(t) = \mathcal{A}x(t) + F(t, x_1, x_2(t)), \quad t \geq t_0, \tag{7.34}$$

in the Hilbert space $\mathcal{H} = W^{1,2}(0, 1) \times L_2(0, 1)$. In the above equation, the infinitesimal operator

$$\mathcal{A} = \begin{bmatrix} 0 & 1 \\ \frac{\partial[a(\xi)\frac{\partial}{\partial \xi}]}{\partial \xi} & + r_2 \frac{\partial}{\partial \xi} & 0 \end{bmatrix} \tag{7.35}$$

possesses the dense domain

$$\mathcal{D}(\mathcal{A}) = \{(x_1, x_2) \in W^{2,2}(0, 1) \times L_2(0, 1) : x_i(0) = 0,$$
$$x_{i_\xi}(1) = -kx_i(1), \ i = 1, 2\} \tag{7.36}$$

and generates a strongly continuous semigroup, whereas the second component

$$F_2(t, x_1, x_2) : \mathbb{R} \times W^{1,2}(0, 1) \times L_2(0, 1) \to L_2(0, 1)$$

of the nonlinear term $F = (0, F_2)$ is defined on solutions $(x_1(\xi, t), x_2(\xi, t))^T$ of (7.34) according to

$$F_2(t, x_1, x_2) = r_0(\xi, t, x_1(\xi, t), x_2(\xi, t))x_2(\xi, t)$$
$$+ r_1(\xi, t, x_1(\xi, t), x_2(\xi, t))x_{1\xi}(\xi, t). \tag{7.37}$$

Since r_0 and r_1 are Lipschitz continuous, the following Lipschitz condition

$$\|F_2(t_1, x_{11}, x_{12}) - F_2(t_2, x_{21}, x_{22})\|_{L_2}$$
$$\leq L[|t_1 - t_2| + \|x_{11} - x_{21}\|_{W^{1,2}} + \|x_{21} - x_{22}\|_{L_2}] \tag{7.38}$$

holds locally in $(t_i, x_{i1}, x_{i2}) \in \mathbb{R} \times W^{1,2}(0,1) \times L_2(0,1)$, $i = 1, 2$, with some generic constant $L > 0$, and hence, a unique strong solution of (7.34), initialized with $(x_1(t_0), x_2(t_0)) \in W^{1,2}(0,1) \times L_2(0,1)$, $x_i(0) = 0$, $x_{i_\xi}(1) = -kx_i(1)$ $i = 1, 2$, turns out to locally exist (see, e.g., [96, p. 187, Theorem 1.5]). Thus, there exists a unique local strong solution to the boundary-value problem (7.31), (7.32) for an arbitrary initial condition

$$z(\xi, t_0) = \phi(\xi) \in W^{1,2}(0,1) : \phi(0) = 0,$$
$$\phi_\xi(1) = -k\phi(1),$$
$$z_t(\xi, t_0) = \phi_1(\xi) \in L_2(0,1) : \phi_1(0)) = 0,$$
$$\phi_{1_\xi}(1) = -k\phi_1(1). \quad (7.39)$$

As in the heat equation case, only strong solutions of (7.31), (7.32), (7.39) are under study.

It should be noted that once specified with $k = 0$, $a = 1$, and $r_i = 0$, $i = 0, 1$, the linear system (7.31), (7.32) generates oscillating solutions and is therefore asymptotically unstable. The aim is to carry out exponential stability conditions for the uncertain nonsmooth system (7.31), (7.32) with $k > 0$.

7.2.1 Exponential Stability

On solutions of (7.31), (7.32), consider the Lyapunov–Krasovskii functional

$$V(z_\xi(\cdot,t), z_t(\cdot,t)) = \int_0^1 [z_\xi \ z_t] P(\xi)[z_\xi \ z_t]^T d\xi, \quad (7.40)$$

proposed in [85], with

$$P(\xi) = \begin{bmatrix} a(\xi)p & \chi\xi \\ \chi\xi & p \end{bmatrix} \quad (7.41)$$

and some constants $p > 0$ and $\chi > 0$ such that $P(\xi)$ is positive definite for all $\xi \in [0, 1]$. Since

$$\begin{bmatrix} a(\xi)p & \chi\xi \\ \chi\xi & p \end{bmatrix} \geq P_1(\xi) \triangleq \begin{bmatrix} a_1 p & \chi\xi \\ \chi\xi & p \end{bmatrix} \text{ for all } \xi \in [0, 1],$$

whereas $P_1(\xi)$ is positive definite for all $\xi \in [0, 1]$ whenever

$$\begin{bmatrix} a_1 p & \chi \\ \chi & p \end{bmatrix} > 0, \quad (7.42)$$

the latter inequality yields the parameter subordination $a_1 p^2 > \chi^2$ that ensures the positive definiteness of $P(\xi)$ for all $\xi \in [0, 1]$.

7.2 Boundary Stabilization of a Semilinear Hyperbolic Equation

If the inequality $\frac{d}{dt}V + 2\delta V \leq 0$ holds on solutions of the boundary-value problem (7.31), (7.32), then by the comparison principle argument [71], it would follow that

$$V(z_\xi(\cdot,t),z_t(\cdot,t)) \leq V(z_\xi(\cdot,t_0),z_t(\cdot,t_0))e^{-2\delta(t-t_0)}. \tag{7.43}$$

Due to (7.33) and (7.42), one has $0 < mI < P(\xi) < MI$ for $\xi \in [0,1]$ and some scalars $0 < m < M$. Therefore, the solution of (7.31), (7.32), initialized with (7.39), would satisfy the bound

$$\int_0^1 [z_\xi^2(\xi,t) + z_t^2(\xi,t)]d\xi \leq \frac{M}{m}e^{-2\delta(t-t_0)}\int_0^1 [\phi_\xi^2(\xi) + \phi_1^2(\xi)]d\xi \tag{7.44}$$

and would be globally continuable to the right (see, e.g., [74, Theorem 23.5]).

For later use, let's derive

$$\frac{d}{dt}\left(2\int_0^1 \xi z_t z_\xi d\xi\right) = 2\int_0^1 \xi z_{tt} z_\xi d\xi + 2\int_0^1 \xi z_t z_{\xi t}d\xi$$
$$= 2\int_0^1 \xi \frac{\partial}{\partial \xi}[az_\xi(\xi,t)]z_\xi d\xi + 2\int_0^1 \xi z_t z_{\xi t}d\xi$$
$$+ 2\int_0^1 \xi[r_0 z_t(\xi,t) + r_1 z_\xi(\xi,t)]z_\xi d\xi$$
$$= \int_0^1 \frac{1}{a}\xi \frac{\partial}{\partial \xi}(az_\xi)^2 d\xi + 2\int_0^1 \xi z_t z_{\xi t}d\xi$$
$$+ 2\int_0^1 \xi[r_0 z_t(\xi,t) + r_1 z_\xi(\xi,t)]z_\xi d\xi$$
$$= -\int_0^1 az_\xi^2 d\xi + a_{|\xi=1}z_\xi^2(1,t) + 2\int_0^1 \xi z_t(\xi,t)z_{\xi t}d\xi$$
$$+ 2\int_0^1 \xi[r_0 z_t(\xi,t) + r_1 z_\xi(\xi,t)]z_\xi d\xi.$$

After integrating by parts, we see that

$$2\int_0^1 \xi z_t z_{\xi t}d\xi = -2\int_0^1 \xi z_{\xi t} z_t d\xi - 2\int_0^1 z_t^2 d\xi + 2z_t^2(1,t).$$

Therefore, $2\int_0^1 \xi z_t z_{\xi t}d\xi = -\int_0^1 z_t^2 d\xi + z_t^2(1,t)$, which results in

$$\frac{d}{dt}\left(2\int_0^1 \xi z_t z_\xi d\xi\right) = -\int_0^1 (z_t^2 + az_\xi^2)d\xi + z_t^2(1,t)$$
$$+ a_{|\xi=1}z_\xi^2(1,t) + 2\int_0^1 \xi[r_0 z_t(\xi,t) + r_1 z_\xi(\xi,t)]z_\xi d\xi.$$

Thus, differentiating V along (7.31), (7.32), we obtain

$$\frac{d}{dt}V + 2\delta V \leq 2p\int_0^1 az_\xi(\xi,t)z_{t\xi}(\xi,t)d\xi$$
$$+ 2p\int_0^1 z_t(\xi,t)z_{tt}(\xi,t)d\xi + \frac{d}{dt}\left(2\chi\int_0^1 \xi z_t z_\xi\right)$$
$$+ \int_0^1 2\delta[apz_\xi^2(\xi,t) + 2\chi\xi z_\xi(\xi,t)z_t^2(\xi,t) + pz_t^2(\xi,t)]d\xi$$
$$= 2p\int_0^1 \left[az_\xi(\xi,t)z_{t\xi}(\xi,t) + z_t(\xi,t)\frac{\partial}{\partial \xi}[az_\xi(\xi,t)]\right]d\xi d\xi$$
$$+ 2p\int_0^1 z_t(\xi,t)[r_0 z_t(\xi,t) + r_1 z_\xi(\xi,t)]d\xi$$
$$+ \chi\bigg[-\int_0^1 (z_t^2 + az_\xi^2)d\xi + z_t^2(1,t) + a_{|\xi=1}k^2 z_t^2(1,t)$$
$$+ 2\int_0^1 \xi[r_0 z_t(\xi,t) + r_1 z_\xi(\xi,t)]z_\xi d\xi\bigg]$$
$$+ \int_0^1 2\delta[az_\xi^2(\xi,t) + 2\chi\xi z_\xi(\xi,t)z_t(\xi,t) + z_t^2(\xi,t)]d\xi.$$

Now integrating by parts and taking into account (7.31) and (7.32) yield

$$\frac{d}{dt}V + 2\delta V \leq -2a_{|\xi=1}kpz_t^2(1,t)$$
$$+2p\int_0^1 z_t(\xi,t)[r_0 z_t(\xi,t) + r_1 z_\xi(\xi,t)]d\xi$$
$$+\chi\Big[-\int_0^1 (z_t^2 + az_\xi^2)d\xi + (1+a_{|\xi=1}k^2)z_t^2(1,t)\quad\quad(7.45)$$
$$+2\int_0^1 \xi[r_0 z_t(\xi,t) + r_1 z_\xi(\xi,t)]z_\xi d\xi\Big]$$
$$+\int_0^1 2\delta[apz_\xi^2(\xi,t) + 2\chi\xi z_\xi(\xi,t)z_t(\xi,t) + pz_t^2(\xi,t)]d\xi.$$

Taking into account that

$$2\int_0^1 \chi\xi[r_0 z_t(\xi,t) + r_1 z_\xi(\xi,t)]z_\xi(\xi,t)d\xi$$
$$\leq \int_0^1 \xi\Big[\frac{\chi^2\beta_0^2}{s_0}z_t^2(\xi,t) + s_0 + 2\chi\beta_1)z_\xi^2(\xi,t)\Big]d\xi$$

for some $s_0 > 0$, setting $\zeta^T(\xi,t) = [z_t(1,t)\ z_\xi(\xi,t)\ z_t(\xi,t)]$, and using $a \geq a_1$, we conclude that

$$\frac{d}{dt}V + 2\delta V \leq \int_0^1 \zeta^T(\xi,t)\Psi\zeta(\xi,t)d\xi \leq 0$$

if

$$\Psi = \begin{bmatrix} \psi_1 & 0 & 0 \\ * & \psi_2 & 2\chi\delta\xi + pr_1 \\ * & * & \psi_3 + \frac{\chi^2\beta_0^2}{s_0}\xi \end{bmatrix} \leq 0, \quad\quad(7.46)$$

where

$$\psi_1 = -2a_1 kp + (1+a_1 k^2)\chi,$$
$$\psi_2 = -a_1\chi + 2\delta a_1 p + s_0 \xi + 2\chi\xi\beta_1,\quad\quad(7.47)$$
$$\psi_3 = -\chi + 2p\beta_0 + 2\delta p.$$

By Schur complements, (7.46) holds if

$$\begin{bmatrix} \psi_1 & 0 & 0 & 0 \\ * & \psi_2 & 2\chi\delta\xi + pr_1 & 0 \\ * & * & \psi_3 & \beta_0\chi\xi \\ * & * & * & -s_0\xi \end{bmatrix} \leq 0. \quad\quad(7.48)$$

It is worth noticing that given k, (7.48) is an LMI, which is affine in $\xi \in [0,1]$, $r_1 \in [-\beta_1, \beta_1]$. Therefore, the LMI (7.48) is feasible if the following LMIs in the four vertices are feasible:

$$\begin{bmatrix} \psi_1 & 0 & 0 & 0 \\ * & \psi_2^{(j)} & 2\chi\delta\xi^{(j)} + pr_1^{(l)} & 0 \\ * & * & \psi_3 & \beta_0\chi\xi^{(j)} \\ * & * & * & -s_0 \end{bmatrix} \leq 0,\ j=1,2;\ l=1,2;$$
$$\psi_2^{(j)} = -a_1\chi + 2\delta a_1 p + s_0 + 2\chi\xi^{(j)}\beta_1,\quad\quad(7.49)$$
$$r_1^{(1)} = \beta_1,\ r_1^{(2)} = -\beta_1,\ \xi^{(1)} = 0,\ \xi^{(2)} = 1.$$

7.2 Boundary Stabilization of a Semilinear Hyperbolic Equation

Note that for the stability analysis, $p = 1$ can be chosen. Moreover, the feasibility of (7.49) implies $\psi_3 \leq 0$ and, thus, $\chi > 0$. Summarizing, the following is obtained.

Theorem 31. *Given $k > 0$ and $\delta > 0$, let the LMIs (7.42) and (7.49) with notation (7.47) and $p = 1$ hold for some χ and s_0. Then a unique strong solution of the boundary-value problem (7.31), (7.32), (7.39) is globally continuable to the right and satisfies (7.44) for all $t \geq t_0$.*

The $\mathcal{H}_\infty$-boundary synthesis is presented next.

7.2.2 $\mathcal{H}_\infty$-Boundary Control

In addition to the hyperbolic equation (7.31), let's now consider its perturbed version

$$\begin{aligned} z_{tt}(\xi,t) &= \tfrac{\partial}{\partial \xi}[az_\xi] + r_0 z_t(\xi,t) + r_1 z_\xi(\xi,t) \\ &\quad + bw(\xi,t),\ t \geq 0,\ 0 \leq \xi \leq 1, \end{aligned} \quad (7.50)$$

where $w(\xi,t) \in L_2(0,\infty; L_2(0,1))$ is an external disturbance; $b = b(\xi,t,z)$ is a Lipschitz continuous function, which is assumed to be uniformly bounded, that is, $|b(\xi,t,z)| \leq b_1$ for all $(\xi,t,z) \in [0,1] \times \mathbb{R}^2$ and some $b_1 > 0$. While internally stabilizing the hyperbolic process, the influence of the admissible external disturbance on the controlled output

$$\bar{z}(\xi,t) = [\alpha z(\xi,t) \quad \bar{\alpha} z_t(\xi,t) \quad d(t,z_t(1,t))u(t)]^T \quad (7.51)$$

is to be attenuated through the boundary actuation at $\xi = 1$:

$$z(0,t) = 0, \quad z_\xi(1,t) = u(t),\ t \geq t_0. \quad (7.52)$$

Hereinafter, $u(t)$ is the control input, d and $\alpha = \alpha(\xi,t,z)$, $\bar{\alpha} = \bar{\alpha}(\xi,t,z)$ are continuous functions, which are uniformly bounded, namely,

$$|\alpha(\xi,t,z,z_t)| \leq \alpha_0,\ |\bar{\alpha}(\xi,t,z,z_t)| \leq \alpha_1,\ |d(t,z)| \leq d_1,$$

for all $(\xi,t,z,z_t) \in [0,1] \times \mathbb{R}^3$, where $\alpha_i \geq 0$, $i = 0,1$ and $d_1 \geq 0$ are some constants. Collocated sensing $y(t) = z_t(1,t)$ at the boundary $\xi = 1$ is the only available information on the process.

The $\mathcal{H}_\infty$-control problem of interest is stated as follows. Given $\gamma > 0$, find a linear static output feedback

$$u(t) = -k z_t(1,t) \quad (7.53)$$

that exponentially stabilizes the unperturbed system (7.31), (7.32) and leads to a negative performance index

$$J = \int_{t_0}^{\infty} \int_0^1 [\bar{z}^T(\xi,t)\bar{z}(\xi,t) - \gamma^2 w^2(\xi,t)]d\xi dt < 0 \quad (7.54)$$

for all the solutions of (7.50), (7.52), initialized with the zero data $z(\xi,t_0) = z_t(\xi,t_0) = 0$, and for all admissible external disturbances $0 \neq w(\xi,t) \in L_2(0,\infty;L_2(0,1))$, under which these solutions are globally continuable to the right.

To solve the stated problem, let's find conditions that guarantee the inequality

$$W(t) \triangleq \frac{d}{dt}V + \int_0^1 [\bar{z}^T(\xi,t)\bar{z}(\xi,t) - \gamma^2 w^2(\xi,t)]d\xi < 0, \quad (7.55)$$

where V is given by (7.40) and the temporal derivative is computed along the closed-loop system (7.50), (7.52). First, by employing Wirtinger's inequality (2.1), we obtain the following:

$$\int_0^1 \bar{z}^T(\xi,t)\bar{z}(\xi,t)d\xi \leq \int_0^1 \left[\alpha_0^2 z^2(\xi,t) + \alpha_1^2 z_t^2(\xi,t) \right.$$
$$\left. + d_1^2 k^2 z_t^2(1,t)\right]d\xi \leq \int_0^1 \left[\frac{1}{2}\alpha_0^2 z_\xi^2(\xi,t) + \alpha_1^2 z_t^2(\xi,t) \right.$$
$$\left. + d_1^2 k^2 z_t^2(1,t)\right]d\xi.$$

Then, analogously to (7.45), one has

$$\frac{d}{dt}V \leq -2a_{|\xi=1}kpz_t^2(1,t)$$
$$+2p\int_0^1 z_t(\xi,t)[r_0 z_t(\xi,t) + r_1 z_\xi(\xi,t) + bw]d\xi$$
$$+\chi\left[-\int_0^1 (z_t^2 + az_\xi^2)d\xi + (1 + a_{|\xi=1}k^2)z_t^2(1,t) \right.$$
$$\left. +2\int_0^1 \xi[r_0 z_t(\xi,t) + r_1 z_\xi(\xi,t) + bw]z_\xi d\xi\right].$$

Along with this, we derive the inequality

$$2\int_0^1 \chi\xi[r_0 z_t(\xi,t) + r_1 z_\xi(\xi,t) + bw]z_\xi(\xi,t)d\xi$$
$$\leq \int_0^1 \left[\frac{\chi^2 \beta_0^2}{s_0}z_t^2(\xi,t) + \frac{\chi^2 b_1^2}{s_1}w^2 \right.$$
$$\left. +(s_0 + s_1 + 2\chi\beta_1)z_\xi^2(\xi,t)\right]d\xi$$

for some $s_0 > 0$ and $s_1 > 0$, Finally, by taking into account $a \geq a_1$, we conclude that

$$W = \frac{d}{dt}V + \int_0^1 [\bar{z}^T\bar{z} - \gamma^2 w^2]d\xi \leq \bar{\zeta}^T \Psi_\gamma \bar{\zeta},$$

7.2 Boundary Stabilization of a Semilinear Hyperbolic Equation

where

$$\bar{\zeta}^T = [z_t(1,t)\ z_\xi(\xi,t)\ z_t(\xi,t)\ w(\xi,t)],$$

$$\Psi_\gamma = \begin{bmatrix} \psi_1 + d_1^2 k^2 & 0 & 0 & 0 \\ * & \psi_{2\gamma} & pr_1 & 0 \\ * & * & \psi_{3\gamma} + \frac{\beta_0^2 \chi^2}{s_0} & pb \\ * & * & * & -\gamma^2 + \frac{b_1^2 \chi^2}{s_1} \end{bmatrix},$$

and

$$\begin{aligned} \psi_1 &= -2a_1 kp + (1 + a_1 k^2)\chi, \\ \psi_{2\gamma} &= -a_1\chi + s_0 + s_1 + \tfrac{1}{2}\alpha_0^2 + 2\chi\beta_1, \\ \psi_{3\gamma} &= -\chi + 2p\beta_0 + \alpha_1^2. \end{aligned} \qquad (7.56)$$

Hence, $W < 0$ if $\Psi_\gamma < 0$, namely, by Schur complements, if

$$\begin{bmatrix} \psi_1 + d_1^2 k^2 & 0 & 0 & 0 & 0 & 0 \\ * & \psi_{2\gamma} & pr_1 & 0 & 0 & 0 \\ * & * & \psi_{3\gamma} & pb & \beta_0\chi & 0 \\ * & * & * & -\gamma^2 & 0 & b_1\chi \\ * & * & * & * & -s_0 & 0 \\ * & * & * & * & * & -s_1 \end{bmatrix} < 0. \qquad (7.57)$$

The LMI (7.57) is affine in $r_1 \in [-\beta_1, \beta_1]$, and $b \in [-b_1, b_1]$. Therefore, it is feasible if it holds in the vertices $r_1 = \pm\beta_1$ and $b = \pm b_1$. Clearly, the four LMIs in the vertices are equivalent to the following one:

$$\begin{bmatrix} \psi_1 + d_1^2 k^2 & 0 & 0 & 0 & 0 & 0 \\ * & \psi_{2\gamma} & p\beta_1 & 0 & 0 & 0 \\ * & * & \psi_{3\gamma} & pb_1 & \beta_0\chi & 0 \\ * & * & * & -\gamma^2 & 0 & b_1\chi \\ * & * & * & * & -s_0 & 0 \\ * & * & * & * & * & -s_1 \end{bmatrix} < 0. \qquad (7.58)$$

Note that if (7.58) is feasible, then the LMI (7.49), used in Theorem 31, holds with a sufficiently small $\delta > 0$. The following result is thus proved.

Theorem 32. *Consider the perturbed input–output system (7.50)–(7.52) with the assumptions above. Given $\gamma > 0$ and $k > 0$, let there exist $p > 0$, χ, s_0, and s_1 such that the LMIs (7.42) and (7.58) are satisfied with the notations given by (7.56). Then the static output feedback (7.53) internally exponentially stabilizes the boundary-value problem (7.50), (7.52) and attenuates the admissible disturbances $w(\xi, t) \in L_2(0, \infty; L_2(0,1))$ in the sense of (7.54).*

To illustrate the proposed synthesis, we invoke a simple example.

Example 6. Consider (7.50)–(7.52) with

$$a_1 = 2,\ \beta_0 = 0.2,\ \beta_1 = 0.3,\ \alpha_0 = \alpha_1 = 1,\ d_1 = 0.1.$$

As mentioned before, the open-loop system is unstable. If we use MATLAB®'s LMI toolbox, the LMIs (7.42) and (7.58) are verified and the static output feedback (7.53) with $k = 1$ is selected according to Theorem 32 to exponentially stabilize the disturbance-free system and to impose the disturbance attenuation level $\gamma = 4.3$ on the perturbed system. By verifying (7.49) in the four vertices, we establish the internal exponential stability of the resulting closed-loop system with the decay rate $\delta = 0.13$.

Part III
Benchmark Applications

Performance and robustness issues of the developed synthesis are experimentally tested in engineering applications to frictional mechanical manipulators, to servomotors with backlash, to underactuated helicopter prototypes, all with incomplete measurements, as well as to the state-feedback stabilization of the current one-dimensional radial profiles in tokamak plasmas. Capabilities of the local nonsmooth $\mathcal{H}_\infty$ synthesis of accounting for hard-to-model friction forces and backlash effects are investigated in experimental studies made for the position-feedback regulation of frictional manipulators and for the output regulation of a servomechanism with backlash. Both fully actuated and underactuated manipulators (a three-link robot manipulator and Pendubot, respectively) are involved to demonstrate attractive features of the approach for regulation and tracking over continuous position measurements, whereas a simple one-link manipulator (inverted pendulum) is invoked to additionally illustrate the efficacy of the approach over sampled-data measurements. Since the absence of full actuation imposes limitations on planning feasible trajectories (as opposed to fully actuated manipulators, which are the case with feedback linearization [115]), the presentation of the so-called virtual constraint approach from [112] precedes the periodic tracking synthesis to be developed for a 3-DOF helicopter prototype of one degree of underactuation. The key idea of the synthesis in question is to carry out an invariant set of geometric relations among the degrees of freedom that could be enforced via the $\mathcal{H}_\infty$-feedback design, resulting in stable limit cycles of interest. Finally, the tokamak plant simulator METIS [9] is utilized to support the LMI-based $\mathcal{H}_\infty$ synthesis, developed in the PDE setting.

Chapter 8
Advanced $\mathcal{H}_\infty$ Synthesis of Fully Actuated Robot Manipulators with Frictional Joints

Motivated by the presence of nonsmooth phenomena caused by dry friction forces, in this chapter we undertake local nonsmooth $\mathcal{H}_\infty$ synthesis of a multilink manipulator with frictional joints. The manipulator is required to follow a desired trajectory and particularly to move from an initial position to a desired one. Since in robotic applications velocity sensors are often omitted to save considerably in cost, volume, and weight [72], the position is assumed to be the only available measurement on the system.

To begin, we review well-known static and dynamic friction models. For certainty, the frictional forces that occur in the manipulator joints are then represented by the Dahl model augmented with viscous friction. This simplest dynamic model captures all the essential features of the general treatment and allows one to straightforwardly apply the proposed nonsmooth $\mathcal{H}_\infty$ synthesis that proves capable of counting for the nonsmooth terms of the Dahl friction model. Along with the velocity compensator, the resulting nonsmooth $\mathcal{H}_\infty$ synthesis necessarily includes friction compensator design, thereby yielding a $\mathcal{H}_\infty$ controller of a higher order compared to that of the plant.

To avoid having to implement the friction compensator, thus saving in the computational cost and physical volume of the controller, we develop ad hoc an alternative discontinuous $\mathcal{H}_\infty$ design based on the discontinuous static Coulomb model that can be viewed as a singularly perturbed version of the Dahl model with an infinitely large stiffness coefficient. Performance issues of the controllers thus developed are illustrated in an experimental study made for a 3-DOF robot manipulator with frictional joints.

In the experimental study, we conclude a better performance of both the nonsmooth and discontinuous $\mathcal{H}_\infty$ controllers, compared to its smooth version derived via the standard nonlinear $\mathcal{H}_\infty$ approach, based on the observation that larger definition domains of the external attenuated disturbances are provided by the discontinuous and nonsmooth controllers. We should point out that implementation of the discontinuous local $\mathcal{H}_\infty$ controller is of the same level of simplicity as that of the aforementioned nonlinear $\mathcal{H}_\infty$ controller, whereas implementation of the nonsmooth local $\mathcal{H}_\infty$ controller is of a higher complexity level due to the need to implement the dynamic friction compensator.

8.1 Scalar Friction Models

Friction is a natural phenomenon representing the tangential reaction force between two surfaces in contact. Since these reaction forces depend on many factors, such as contact geometry and surface materials, displacement and relative velocities of contacting bodies, and the presence of lubrication, among others, it is hardly possible to deduce a general friction model from physical first principles that would describe the frictional forces that occur in the manipulator joints. Instead, phenomenological models, capturing essential friction features, are brought into play and scalar friction models of interest are reviewed later in this chapter (see the surveys [8, 10] for details and for other existing friction models).

8.1.1 Static Models

The classical friction models are described by static maps between velocity and frictional forces. The main idea of such a model is that friction opposes motion and its magnitude is independent of velocity v and contact area.

The Coulomb friction model

$$F(v) = F_C \operatorname{sign} v \tag{8.1}$$

is used for the dry friction force and is an ideal relay model, multivalued for zero velocity:

$$\operatorname{sign} v = \begin{cases} 1, & \text{if } v > 0, \\ [-1, 1], & \text{if } v = 0, \\ -1, & \text{if } v < 0. \end{cases} \tag{8.2}$$

Since it does not specify the frictional force $F(v)$ for zero velocity, the static force $F(0)$ is admitted to counteract external forces below the Coulomb friction level F_C. Thus, stiction, describing the Coulomb friction force at rest, can take on any value in the segment $[-F_C, F_C]$, thereby yielding the meaning of the corresponding state equation in the sense of Filippov [42].

The viscous friction model

$$F(v) = F_v v \tag{8.3}$$

with a viscous friction coefficient $F_v > 0$ is used for the frictional force caused by the viscosity of lubricants. When combined with the Coulomb friction, it is often modified to

$$F(v) = F_v v + F_C \operatorname{sign} v. \tag{8.4}$$

In order to account for the observed destabilizing Stribeck phenomenon at very low velocities, the latter model can be augmented with the Stribeck friction model $\sigma_s e^{-(v/v_s)^2} \operatorname{sign} v$, where the constants $\sigma_s > 0$ and $v_s > 0$ stand for the Stribeck level and the Stribeck velocity, respectively. The resulting model is then given by

$$F(v) = F_v v + [F_C + \sigma_s e^{-(\frac{v}{v_s})^2}] \operatorname{sign} v, \tag{8.5}$$

where the stiction force level $F_S = \sigma_s + F_C$ is admitted to be higher than the Coulomb level F_C.

8.1.2 Dynamic Models

To better match experimental data, dynamic modeling of friction is typically involved. Proposed by Dahl [36], the dynamic model

$$\dot{F}_D = \sigma_1 v - \sigma_1 |v| \frac{F_D}{F_C}, \tag{8.6}$$

where $\sigma_1 > 0$ and $F_C > 0$ are the stiffness and the Coulomb friction level, respectively, describes the spring-like behavior of the friction force F_D during stiction when the velocity v of the contacting body is infinitesimal.

It is of interest to note that when we formally introduce the parameter $\mu = \sigma_1^{-1}$, the above dynamic model (8.6) specializes to

$$\mu \dot{F}_D = v - |v| \frac{F_D}{F_C}, \tag{8.7}$$

which can be viewed as a singular perturbation of the static Coulomb model (8.1) with a small parameter $\mu > 0$. Indeed, letting $\mu = 0$ in (8.7), one formally arrives at the static Coulomb model (8.1). Since the Dahl model (8.6) is nonsmooth rather than discontinuous, it can be viewed as a regularization of the discontinuous Coulomb model (8.1) as $\mu = \sigma_1^{-1} \to 0$, namely, as $\sigma_1 \to \infty$.

While being essentially Coulomb friction with a lag in the change of the frictional force when the motion direction is changed, the Dahl model (8.6) does not capture the Stribeck effect. In order to account for the Stribeck effect, one can use the LuGre friction model

$$F_L = F_v v + \sigma_1 \eta + \sigma_2 \frac{d\eta}{dt} \tag{8.8}$$

from [27]. In the LuGre model (8.8), the friction interface is thought of as a contact between bristles, $F_v > 0$ is a viscous friction coefficient, $\sigma_1 > 0$ is the stiffness, $\sigma_2 > 0$ is a damping coefficient, η is the average deflection of the bristles, whose dynamics are governed by

$$\frac{d\eta}{dt} = v - \frac{\sigma_1|v|}{F_C + [F_S - F_C]e^{-(\frac{v}{v_s})^2}}\eta, \qquad (8.9)$$

where $F_C > 0$ is the Coulomb friction level, $F_S > 0$ is the level of the stiction force, $v_s > 0$ is the Stribeck velocity, and v is the actual velocity of the contacting body.

Thus, the complete LuGre model (8.8), (8.9) is characterized by six parameters: $F_v, \sigma_1, \sigma_2, F_C, F_S, v_s$. It reduces to the Dahl model (8.6) if $F_v = 0, \sigma_2 = 0$, and $F_S = F_C$. In turn, for steady-state motion when v is constant and $\dot{\eta} = 0$, the relation between the velocity and the LuGre friction force (8.8) is given by the classical model (8.5).

8.2 Problem Statement

A mathematical model for a frictional mechanical manipulator whose links are joined together with revolute joints is given by

$$M(q)\ddot{q} + C(q,\dot{q})\dot{q} + G(q) + F(\dot{q}) = \tau + w_1, \qquad (8.10)$$

where $q \in \mathbb{R}^n$ is a position, $\tau \in \mathbb{R}^n$ is a control input, $w_1 \in \mathbb{R}^n$ is an external disturbance, and $F(\dot{q})$, $G(q)$, $M(q)$, and $C(q,\dot{q})$ are matrix functions of appropriate dimensions. From the physical point of view, q is the vector of generalized coordinates, τ is the vector of external torques, $M(q)$ is the inertia matrix, symmetric and positive definite for all $q \in \mathbb{R}^n$, $C(q,\dot{q})\dot{q}$ is the vector of Coriolis and centrifugal torques, $G(q)$ is the vector of gravitational torques, and the components $F_i(\dot{q}_i)$, $i = 1, \ldots, n$, of $F(\dot{q})$ are the frictional forces acting independently in each joint.

Throughout, the functions $G(q)$, $M(q)$, and $C(q,\dot{q})$ are twice continuously differentiable, whereas the frictional forces are represented as a combination

$$F_i = \sigma_{0i}\dot{q}_i + F_{di}, \ i = 1, \ldots, n, \qquad (8.11)$$

of viscous friction $\sigma_{0i}\dot{q}_i$ and the dry friction F_{di}, governed by the following Dahl dynamic model:

$$\dot{F}_{di} = \sigma_{1i}\dot{q}_i - \sigma_{1i}|\dot{q}_i|\frac{F_{di}}{F_{ci}} + w_{2i}. \qquad (8.12)$$

Incorporating the detailed LuGre friction modeling into the present study is rather technical and is intentionally omitted to experimentally investigate if the influence of nonlinear phenomena such as the Stribeck effect, ignored by the Dahl model, is attenuated by the nonsmooth $\mathcal{H}_\infty$-synthesis.

8.2 Problem Statement

In the above equations, $\sigma_{0i} > 0$, $\sigma_{1i} > 0$, and $F_{ci} > 0$ are the viscous friction coefficient, the stiffness, and the Coulomb friction level, respectively, corresponding to the ith manipulator joint; w_{2i} is an external disturbance that is involved to account for inadequacies of the friction modeling. Clearly, the above componentwise relations can be rewritten in the vector form

$$F = \sigma_0 \dot{q} + F_d, \quad (8.13)$$

$$\dot{F}_d = \sigma_1 \dot{q} - \sigma_1 diag\{|\dot{q}_i|\} F_c^{-1} F_d + w_2, \quad (8.14)$$

where $F = col\{F_i\}$, $F_d = col\{F_{di}\}$, $q = col\{q_i\}$ $\sigma_0 = diag\{\sigma_{0i}\}$, $\sigma_1 = diag\{\sigma_{1i}\}$, $F_c = diag\{F_{ci}\}$, $w_2 = col\{w_{2i}\}$, and the notations $diag$ and col are used to denote a diagonal matrix and a column vector, respectively.

Let $q_d(t) = col\{q_{di}(t)\}$ be a desired twice continuously differentiable trajectory for the robot manipulator to follow such that the functions $q_d(t)$, $\dot{q}_d(t), \ddot{q}_d(t)$ are uniformly bounded in t. Then if there were no initial and external disturbances, the desired motion could be enforced by the external torque

$$\tau_d = M(q_d)\ddot{q}_d + C(q_d, \dot{q}_d)\dot{q}_d + G(q_d) + F_n, \quad (8.15)$$

where $F_n(t) = col\{F_{ni}\}$ is the nominal frictional force

$$F_n = \sigma_0 \dot{q}_d + F_{nd} \quad (8.16)$$

computed along the desired trajectory according to the undisturbed friction model

$$\dot{F}_{nd} = \sigma_1 \dot{q}_d - \sigma_1 diag\{|\dot{q}_{di}|\} F_c^{-1} F_{nd} \quad (8.17)$$

subject to the initial condition

$$F_{nd}(t_0) = 0 \quad (8.18)$$

at some time instant $t_0 \in \mathbb{R}$ [the absence of initial and external disturbances means that $q(t_0) = q_d(t_0)$, $\dot{q}(t_0) = \dot{q}_d(t_0)$, $F_d(t_0) = 0$, and $w_1 = w_2 = 0$].

The objective is to design a controller of the form

$$\tau = \tau_d + u \quad (8.19)$$

that imposes on the disturbance-free manipulator motion the desired stability properties around $q_d(t)$ while also locally attenuating the effect of the disturbances. Thus, the controller to be constructed consists of the trajectory feedforward compensator (8.15) and a disturbance attenuator $u(t)$, internally stabilizing the closed-loop system around the desired trajectory.

For certainty, the present investigation is confined to the position-tracking control problem, where (1) The output to be controlled is given by

$$z = \rho \begin{bmatrix} 0 \\ q - q_d \end{bmatrix} + \begin{bmatrix} 1 \\ 0 \end{bmatrix} u \quad (8.20)$$

with a positive weight coefficient ρ. (2) The position measurements

$$y = q + w_0, \quad (8.21)$$

corrupted by the error vector $w_0(t) \in \mathbb{R}^n$, are only available. The extension to a general case is straightforward.

The $\mathcal{H}_\infty$-position-tracking control problem for robot manipulators with friction can formally be stated as follows. Given a mechanical system (8.10)–(8.21), a desired trajectory $q_d(t)$ to track, and a real number $\gamma > 0$, we are required to find (if any) a causal dynamic feedback controller $u = \mathcal{K}(y,t)$ with internal state $\xi \in \mathbb{R}^s$ such that the undisturbed closed-loop system is uniformly asymptotically stable around $q_d(t)$ and its $\mathcal{L}_2$ gain is locally less than γ: That is, inequality

$$\int_{t_0}^{t_1} \|z(t)\|^2 dt < \gamma^2 \int_{t_0}^{t_1} \|w(t)\|^2 dt \quad (8.22)$$

is satisfied for all $t_1 > t_0$ and all piecewise-continuous functions $w(t) = (w_0(t), w_1(t), w_2(t))^T$ for which the state trajectory of the closed-loop system starting from the initial point $(q(t_0), \dot{q}(t_0), F_d(t_0), \xi(t_0)) = (q_d(t_0), \dot{q}_d(t_0), 0, 0)$ remains in some neighborhood of the desired trajectory $q_d(t)$ for all $t \in [t_0, t_1]$.

In a particular case where the desired trajectory $q_d(t)$ is specified as an equilibrium point $q_d \in \mathbb{R}^n$ of the closed-loop system, the above problem is apparently reduced to a $\mathcal{H}_\infty$-regulation problem, which is stated as follows. Given a mechanical system (8.10)–(8.21), a desired endpoint q_d, and a real number $\gamma > 0$, we are required to find (if any) a controller

$$\tau = G(q_d) + u, \quad (8.23)$$

composed of the gravitational compensation term $G(q_d)$ and a causal dynamic feedback controller $u = \mathcal{K}(y)$ with internal state $\xi \in \mathbb{R}^s$ such that the undisturbed closed-loop system is uniformly asymptotically stable around the endpoint q_d and its $\mathcal{L}_2$ gain is locally less than γ. While stating the regulation problem, we have taken into account that the external torque (8.15), enforcing the undisturbed closed-loop system to possess an equilibrium at the endpoint q_d, consists of the only gravitational term $G(q_d)$ because in the present case, $\dot{q}_d = 0, \ddot{q}_d = 0$, and by virtue of this, the nominal frictional force F_n, governed by (8.16)–(8.18), has no value as well.

8.3 Nonsmooth Synthesis via Dynamic Friction Compensation

We will subsequently study, side by side, the $\mathcal{H}_\infty$-tracking-control problem, where

$$\dot{q}_{d_i}(t) \neq 0 \ for \ almost \ all \ t, i = 1, \ldots n, \quad (8.24)$$

8.3 Nonsmooth Synthesis via Dynamic Friction Compensation

and the $\mathcal{H}_\infty$-regulation problem, where

$$\dot{q}_d(t) \equiv 0. \tag{8.25}$$

To begin, let's introduce the state deviation vector $x = (x_1^T, x_2^T, x_3^T)^T$, where $x_1(t) = q(t) - q_d(t)$ is the position deviation from the desired trajectory $q_d(t)$, $x_2(t) = \dot{q}(t) - \dot{q}_d(t)$ is the velocity deviation from the desired velocity $\dot{q}_d(t)$, and $x_3(t) = F_d(t) - F_{nd}(t)$ is the Dahl friction deviation from that computed according to (8.17), (8.18).

Next, we'll rewrite the state equations (8.10)–(8.21) in terms of these deviations:

$$\dot{x}_1 = x_2,$$
$$\dot{x}_2 = -\ddot{q}_d - M^{-1}(x_1 + q_d)[C(x_1 + q_d, x_2 + \dot{q}_d)(x_2 + \dot{q}_d)$$
$$+ G(x_1 + q_d) - M(q_d)\ddot{q}_d - C(q_d, \dot{q}_d)\dot{q}_d - G(q_d)$$
$$+ \sigma_0 x_2 + x_3 - u - w_1],$$
$$\dot{x}_3 = \sigma_1 x_2 - \sigma_1 diag\{|x_{2i} + \dot{q}_{di}|\} F_c^{-1} x_3$$
$$- \sigma_1 [diag\{|x_{2i} + \dot{q}_{di}|\} - diag\{|\dot{q}_{di}|\}] F_c^{-1} F_{nd} + w_2, \tag{8.26}$$

$$z = \rho \begin{bmatrix} 0 \\ x_1 \end{bmatrix} + \begin{bmatrix} 1 \\ 0 \end{bmatrix} u, \tag{8.27}$$

$$y = x_1 + q_d + w_0. \tag{8.28}$$

Since for almost all t the right-hand sides of the error equations (8.26)–(8.28) subject to (8.24) turn out to be twice continuously differentiable in x locally around the origin $x = 0$, the above $\mathcal{H}_\infty$-tracking control problem is nothing other than the earlier studied nonlinear $\mathcal{H}_\infty$-control problem for the nonsmooth time-varying system (6.1)–(6.3) specified as follows:

$$f(x,t) = f_1(x,t) + f_2(x,t),$$

$$f_1(x,t) = \begin{bmatrix} x_2 \\ -\ddot{q}_d - M^{-1}(x_1 + q_d)[C(x_1 + q_d, x_2 + \dot{q}_d)(x_2 + \dot{q}_d)] \\ \sigma_1 x_2 - \sigma_1 diag\{|x_{2i} + \dot{q}_{di}|\} F_c^{-1} x_3 \end{bmatrix}$$
$$+ \begin{bmatrix} 0 \\ -M^{-1}(x_1 + q_d)[G(x_1 + q_d) - M(q_d)\ddot{q}_d] \\ -\sigma_1 [diag\{|x_{2i} + \dot{q}_{di}|\} - diag\{|\dot{q}_{di}|\}] F_c^{-1} F_{nd}(t) \end{bmatrix}$$
$$+ \begin{bmatrix} 0 \\ M^{-1}(x_1 + q_d)[C(q_d, \dot{q}_d)\dot{q}_d + G(q_d) - \sigma_0 x_2 - x_3] \\ 0 \end{bmatrix},$$

$$f_2(x,t) = 0, \qquad (8.29)$$

$$g_1(x,t) = \begin{bmatrix} 0 & 0 & 0 \\ 0 & M^{-1}(x_1+q_d) & 0 \\ 0 & 0 & I \end{bmatrix},$$

$$g_2(x,t) = \begin{bmatrix} 0 \\ M^{-1}(x_1+q_d) \\ 0 \end{bmatrix},$$

$$h_1(x,t) = \rho \begin{bmatrix} 0 \\ x_1 \end{bmatrix}, \quad h_2(x,t) = x_1 + q_d(t),$$

$$k_{12}(x,t) = \begin{bmatrix} I \\ 0 \end{bmatrix}, \quad k_{21}(x,t) = [\,I\ 0\ 0\,]. \qquad (8.30)$$

Now, by applying Theorems 24 and 25 to system (6.1)–(6.3), thus specified, a local solution of the afore-stated $\mathcal{H}_\infty$-tracking control problem is obtained along the line of the nonsmooth time-varying/time-periodic $\mathcal{H}_\infty$ synthesis of Sect. 6.1.

Theorem 33. *Let (8.24) hold for the desired trajectory to follow and let Conditions C1, C2 of Sect. 6.1 hold for the matrix functions $A(t)$, $B_1(t)$, $B_2(t)$, $C_1(t)$, $C_2(t)$, governed by (6.35), (8.29), and (8.30). Let $(P_\varepsilon(t),\ Z_\varepsilon(t))$ denote the corresponding bounded positive definite solution of (6.40), (6.41) under some $\varepsilon > 0$. Then the output feedback*

$$\dot{\xi} = f(\xi,t) + \left[\frac{1}{\gamma^2}g_1(\xi,t)g_1^T(\xi,t) - g_2(\xi,t)g_2^T(\xi,t)\right]P_\varepsilon(t)\xi$$
$$+ Z_\varepsilon(t)C_2^T(t)[y(t) - h_2(\xi,t)],$$
$$u = -g_2^T(\xi,t)P_\varepsilon(t)\xi \qquad (8.31)$$

subject to (8.29), (8.30) is a local solution of the $\mathcal{H}_\infty$-tracking control problem for the friction mechanical manipulator (8.10)–(8.21). If, in addition, the trajectory $q_d(t)$ is T-periodic and Conditions C1' and C2' of Sect. 6.2 hold for the same matrix functions $A(t)$, $B_1(t)$, $B_2(t)$, $C_1(t)$, $C_2(t)$, which are now T-periodic, then $(P_\varepsilon(t),\ Z_\varepsilon(t))$ is T-periodic, too, and the output feedback (8.31) yields a T-periodic solution of the corresponding $\mathcal{H}_\infty$-tracking control problem.

Proof. Assumptions A1–A5 of Chap. 6 are straightforwardly verified for the time-varying system (6.1)–(6.3) specified through (8.29), (8.30) subject to (8.24). System (6.1)–(6.3) thus specified represents the error equations (8.26)–(8.28) for the friction mechanical manipulator (8.10)–(8.21), and in order to complete the proof, we need to apply Theorem 24 and, respectively, Theorem 25 in the time-periodic case to the error equations (8.26)–(8.28).

8.3 Nonsmooth Synthesis via Dynamic Friction Compensation

It is worth noticing that a local solution of the $\mathcal{H}_\infty$-regulation problem in question cannot straightforwardly be deduced from that of the $\mathcal{H}_\infty$-tracking control problem, because in contrast to the latter problem, the error equations for the former problem contain a nontrivial nonsmooth function $f_2(x,t) \neq 0$ [cf. that of (8.29)]. Indeed, substituting the nominal velocity $\dot{q}_d(t) = 0$ and Dahl friction $F_{nd}(t) = 0$, computed according to (8.17), (8.18), (8.25), into (8.26) yields the autonomous error equations

$$\dot{x}_1 = x_2,$$
$$\dot{x}_2 = -M^{-1}(x_1 + q_d)[C(x_1 + q_d, x_2)x_2$$
$$+ G(x_1 + q_d) - G(q_d) + \sigma_0 x_2 + x_3 - u - w_1],$$
$$\dot{x}_3 = \sigma_1 x_2 - \sigma_1 diag\{|x_{2i}|\} F_c^{-1} x_3 + w_2 \tag{8.32}$$

with respect to the desired position $q = q_d$, nominal velocity $\dot{q} = 0$, and nominal Dahl friction $F_{nd} = 0$. Since the right-hand side of (8.32) is nonsmooth around the origin $x = 0$ for almost all t, relations (8.29), (8.30) for the regulation problem should be modified to

$$f(x) = f_1(x) + f_2(x),$$

$$f_1(x) = \begin{bmatrix} x_2 \\ M^{-1}(x_1 + q_d)[G(q_d) - C(x_1 + q_d, x_2)x_2] \\ \sigma_1 x_2 \end{bmatrix}$$
$$- \begin{bmatrix} 0 \\ M^{-1}(x_1 + q_d)[G(x_1 + q_d) + \sigma_0 x_2 + x_3] \\ 0 \end{bmatrix},$$

$$f_2(x) = \begin{bmatrix} 0 \\ 0 \\ -\sigma_1 diag\{|x_{2i}|\} F_c^{-1} x_3 \end{bmatrix}, \tag{8.33}$$

$$g_1(x) = \begin{bmatrix} 0 & 0 & 0 \\ 0 & M^{-1}(x_1 + q_d) & 0 \\ 0 & 0 & I \end{bmatrix},$$

$$g_2(x) = \begin{bmatrix} 0 \\ M^{-1}(x_1 + q_d) \\ 0 \end{bmatrix},$$

$$h_1(x) = \rho \begin{bmatrix} 0 \\ x_1 \end{bmatrix}, \quad h_2(x) = x_1 + q_d,$$

$$k_{12}(x) = \begin{bmatrix} I \\ 0 \end{bmatrix}, \quad k_{21}(x) = [I\ 0\ 0]. \tag{8.34}$$

Since by virtue of the well-known inequality

$$2|a|b \le a^2 + b^2, \; a, b \in \mathbb{R},$$

the components $f_{2i}(x)$, $i = 1, 2, 3$ of the nonsmooth function $f_2(x)$, given by (8.33), satisfy $f_{2i}(x) \le \sigma \|x\|^2$ for $\sigma = 0.5 \max_i \sigma_{1i} F_{ci}^{-1}$ and all $x \in \mathbb{R}^n$, a local nonsmooth solution of the $\mathcal{H}_\infty$-regulation problem can be derived along the lines of the nonsmooth autonomous $\mathcal{H}_\infty$ synthesis of Sect. 6.2 by applying Theorem 26 to system (6.1)–(6.3) specified with (8.33), (8.34).

Theorem 34. *Let relation (8.25) be in force and let q_d be a desired regulation point. Let Conditions $C1''$, $C2''$ of Sect. 6.2 hold for the time-invariant matrices A, B_1, B_2, C_1, C_2, governed by (6.35), (8.33), (8.34), and let $(P_\varepsilon, Z_\varepsilon)$ denote the corresponding positive definite solution of (6.64), (6.65) under the above matrices A, B_1, B_2, C_1, C_2 and some $\varepsilon > 0$. Then the output feedback*

$$\dot{\xi} = f(\xi) + \left[\frac{1}{\gamma^2} g_1(\xi) g_1^T(\xi) - g_2(\xi) g_2^T(\xi) \right] P_\varepsilon \xi + Z_\varepsilon C_2^T [y(t) - h_2(\xi)],$$

$$u = -g_2^T(\xi) P_\varepsilon \xi, \tag{8.35}$$

specified with (8.33), (8.34), is a local solution of the $\mathcal{H}_\infty$-position regulation problem in question.

Proof. Since, by inspection, Assumptions A1–A5 of Chap. 6 hold for the time-invariant system (8.32), which represents the manipulator equations (8.10)–(8.21) in terms of the state deviation with respect to the desired position q_d, one may conclude the validity of Theorem 34 by applying Theorem 26.

8.4 Discontinuous Synthesis via Static Friction Compensation

In the sequel, the proposed nonsmooth $\mathcal{H}_\infty$ synthesis is reduced to the discontinuous one provided that the dry friction forces, occurring in the joints of the robot manipulator, are governed by the static Coulomb model and the dynamic friction compensation is therefore no longer required.

As mentioned in Sect. 8.1.2, the dynamic Dahl model (8.6) reduces to the static Coulomb model (8.1) as $\sigma_1 \to \infty$. Then, if we let $\mu = \sigma_1^{-1} \to 0$, the nominal frictional force (8.16) approaches the static one

$$F_n = \sigma_0 \dot{q}_d + F_c \operatorname{sign} \dot{q}_d, \tag{8.36}$$

and the external torque (8.15), enforcing the undisturbed closed-loop system to follow a desired trajectory $q_d(t)$, specializes to

$$\tau_d = M(q_d) \ddot{q}_d + C(q_d, \dot{q}_d) \dot{q}_d + G(q_d) + \sigma_0 \dot{q}_d + F_c \operatorname{sign} \dot{q}_d. \tag{8.37}$$

8.4 Discontinuous Synthesis via Static Friction Compensation

Furthermore, the third equation of (8.26) is represented in the form

$$\mu \dot{x}_3 = x_2 - diag\{|x_{2i} + \dot{q}_{di}|\} F_c^{-1} x_3$$
$$-[diag\{|x_{2i} + \dot{q}_{di}|\} - diag\{|\dot{q}_{di}|\}] F_c^{-1} F_c \text{sign } \dot{q}_d, +\mu w_2 \quad (8.38)$$

resulting in the singular perturbation of the overall system (8.26) with a small parameter $\mu > 0$. Then letting $\mu = 0$ in (8.38) yields

$$x_3 = F_c \text{sign } (x_2 + \dot{q}_d) - F_c \text{sign } \dot{q}_d, \quad (8.39)$$

and by substituting the right-hand side of (8.39) into (8.26), one arrives at the so-called slow dynamics, governed by

$$\dot{x}_1 = x_2,$$
$$\dot{x}_2 = -\ddot{q}_d - M^{-1}(x_1 + q_d)[C(x_1 + q_d, x_2 + \dot{q}_d)(x_2 + \dot{q}_d)$$
$$+ G(x_1 + q_d) - M(q_d)\ddot{q}_d - C(q_d, \dot{q}_d)\dot{q}_d - G(q_d)$$
$$+ \sigma_0 x_2 + F_c \text{sign } (x_2 + \dot{q}_d) - F_c \text{sign } \dot{q}_d - u - w_1]. \quad (8.40)$$

As opposed to system (8.26), the state vector $(x_1, x_2)^T$ of the reduced system (8.40) no longer involves the friction components and the disturbance vector $w = (w_0, w_1)^T$ consists of two components only. With this in mind, the $\mathcal{H}_\infty$-control problem for system (8.40) is stated in the manner similar to that of Sect. 8.2 and it is readily given in terms of the generic system (6.1)–(6.3) if the latter system is specified with

$$f(x,t) = f_1(x,t) + f_2(x,t),$$
$$f_1(x,t) = \begin{bmatrix} x_2 \\ M^{-1}(x_1 + q_d)[G(q_d) - C(x_1 + q_d, x_2)x_2] \end{bmatrix}$$
$$- \begin{bmatrix} 0 \\ M^{-1}(x_1 + q_d)[G(x_1 + q_d) + \sigma_0 x_2] \end{bmatrix}$$
$$- \begin{bmatrix} 0 \\ M^{-1}(x_1 + q_d) F_c [\text{sign } (x_2 + \dot{q}_d) - \text{sign } \dot{q}_d] \end{bmatrix},$$
$$f_2(x,t) = 0, \quad (8.41)$$
$$g_1(x,t) = \begin{bmatrix} 0 & 0 \\ 0 & M^{-1}(x_1 + q_d) \end{bmatrix},$$
$$g_2(x,t) = \begin{bmatrix} 0 \\ M^{-1}(x_1 + q_d) \end{bmatrix},$$

$$h_1(x,t) = \rho \begin{bmatrix} 0 \\ x_1 \end{bmatrix}, \quad h_2(x,t) = x_1 + q_d,$$

$$k_{12}(x,t) = \begin{bmatrix} I \\ 0 \end{bmatrix}, \quad k_{21}(x,t) = [\, I \; 0 \,], \tag{8.42}$$

where for almost all t, the function $f_1(x,t)$ is twice continuously differentiable in x locally around the origin $x = 0$ provided that (8.24) holds for the desired trajectory $q_d(t)$ to be tracked.

The following result, specifying the nonsmooth time-varying/time-periodic $\mathcal{H}_\infty$ synthesis of Sect. 6.1 to the present case, is then in order.

Theorem 35. *Let (8.24) hold for the desired trajectory to follow and let Conditions C1, C2 of Sect. 6.1 hold for the matrix functions $A(t)$, $B_1(t)$, $B_2(t)$, $C_1(t)$, $C_2(t)$, governed by (6.35), (8.41), (8.42). Let $(P_\varepsilon(t), Z_\varepsilon(t))$ denote the corresponding bounded positive definite solution of (6.40), (6.41) under some $\varepsilon > 0$. Then the output feedback*

$$\dot{\xi} = f(\xi,t) + \left[\frac{1}{\gamma^2} g_1(\xi,t) g_1^T(\xi,t) - g_2(\xi,t) g_2^T(\xi,t)\right] P_\varepsilon(t)\xi$$

$$+ Z_\varepsilon(t) C_2^T(t)[y(t) - h_2(\xi,)],$$

$$u = -g_2^T(\xi,t) P_\varepsilon(t)\xi \tag{8.43}$$

subject to (8.41), (8.42) is a local solution of the $\mathcal{H}_\infty$-tracking control problem for the frictional mechanical manipulator (8.10), governed by the classical static friction model (8.4) and driven by (8.19), (8.37), (8.43). If, in addition, the trajectory $q_d(t)$ is T-periodic and Conditions $C1'$ and $C2'$ of Sect. 6.2 hold for the same matrix functions $A(t)$, $B_1(t)$, $B_2(t)$, $C_1(t)$, $C_2(t)$, which are now T-periodic, then $(P_\varepsilon(t), Z_\varepsilon(t))$ is T-periodic, too, and the output feedback (8.43) yields a T-periodic solution of the corresponding $\mathcal{H}_\infty$-tracking control problem.

Proof. As in the proof of Theorem 33, Assumptions A1–A5 of Chap. 6 are straightforwardly verified for the time-varying system (6.1)–(6.3) specified with (8.41), (8.42) subject to (8.24). By applying Theorems 24 and 25 to system (6.1)–(6.3), thus specified, we've established Theorem 35.

For the $\mathcal{H}_\infty$-regulation problem where relation (8.25) holds, setting $\mu = \sigma_1^{-1}$ yields the third equation of (8.32) in the form

$$\mu \dot{x}_3 = x_2 - diag\{|x_{2i}|\} F_c^{-1} x_3 - \mu w_2. \tag{8.44}$$

The resulting system (8.32) is thus singularly perturbed with a small positive parameter μ, and by substituting $\mu = 0$ in (8.44), we formally obtain the so-called slow dynamics

8.4 Discontinuous Synthesis via Static Friction Compensation

$$\dot{x}_1 = x_2,$$
$$\dot{x}_2 = -M^{-1}(x_1 + q_d)[C(x_1 + q_d, x_2)x_2 + G(x_1 + q_d)$$
$$-G(q_d) + \sigma_0 x_2 + F_c \text{sign } x_2 - u - w_1], \quad (8.45)$$

with the dry friction forces governed by the static Coulomb model

$$x_3 = F_c \text{sign } x_2. \quad (8.46)$$

Taking into account that the reduced system (8.45) no longer involves the friction components and the disturbance vector $w = (w_0, w_1)^T$ consists of two components only, we state the $\mathcal{H}_\infty$-regulation problem for system (8.45) in a manner similar to that of Sect. 8.2. It is readily given in terms of the generic system (6.1)–(6.3) if the latter system is specified with

$$f(x) = f_1(x) + f_2(x),$$

$$f_1(x) = \begin{bmatrix} x_2 \\ M^{-1}(x_1 + q_d)[G(q_d) - C(x_1 + q_d, x_2)x_2] \end{bmatrix}$$
$$- \begin{bmatrix} 0 \\ M^{-1}(x_1 + q_d)[G(x_1 + q_d) + \sigma_0 x_2] \end{bmatrix},$$

$$f_2(x) = \begin{bmatrix} 0 \\ -M^{-1}(x_1 + q_d)F_c \text{sign } x_2 \end{bmatrix}, \quad (8.47)$$

$$g_1(x) = \begin{bmatrix} 0 & 0 \\ 0 & M^{-1}(x_1 + q_d) \end{bmatrix},$$

$$g_2(x) = \begin{bmatrix} 0 \\ M^{-1}(x_1 + q_d) \end{bmatrix},$$

$$h_1(x) = \rho \begin{bmatrix} 0 \\ x_1 \end{bmatrix}, \quad h_2(x) = x_1 + q_d,$$

$$k_{12}(x) = \begin{bmatrix} I \\ 0 \end{bmatrix}, \quad k_{21}(x) = [\,I\ 0\,]. \quad (8.48)$$

It should be noted that the $\mathcal{H}_\infty$-synthesis procedure, resulting from Theorem 26, is not applicable to system (6.1)–(6.3) thus specified, because the extracted Coulomb friction term $f_2(x)$ in (8.47) is not simply nonsmooth, but it is discontinuous in the point of interest. It is expected, however, that the $\mathcal{H}_\infty$ synthesis of the discontinuous system (6.1)–(6.3), which corresponds to the plant equations (8.45) with the Coulomb friction model (8.46), namely, specified with (8.47), (8.48), can be derived from the one corresponding to the plant equations with the Dahl friction model, that is, specified with (8.33), (8.34), by extracting the corresponding slow dynamics from the dynamic output feedback (8.35).

By applying the same line of reasoning as that used in deriving the slow dynamics (8.45) of (8.32) as $\mu = \sigma^{-1} \to \infty$, we see that the following nonsmooth autonomous $\mathcal{H}_\infty$ synthesis, extracted from that of Sect. 6.2, appears to be in force.

Proposition 1. *Let Conditions $C1''$ and $C2''$ of Sect. 6.2 hold for the matrix functions A, B_1, B_2, C_1, C_2, governed by (6.35), (8.47), (8.48), and let $(P_\varepsilon, Z_\varepsilon)$ be the corresponding positive definite solution of (6.64), (6.65) under the matrix functions A, B_1, B_2, C_1, C_2 thus specified and some $\varepsilon > 0$. Then the output feedback*

$$\dot{\xi} = f(\xi) + \left[\frac{1}{\gamma^2}g_1(\xi)g_1^T(\xi) - g_2(\xi)g_2^T(\xi)\right]P_\varepsilon\xi + Z_\varepsilon C_2^T[y(t) - h_2(\xi)],$$
$$u = -g_2^T(\xi)P_\varepsilon\xi, \tag{8.49}$$

specified with (8.47), (8.48), is a local solution of the $\mathcal{H}_\infty$-position regulation problem.

Proof. Since the Coulomb friction term $f_2(x)$ in (8.47) is discontinuous, the proof of Proposition 1 is not as straightforward as that of Theorem 34 and only experimental evidence of the validity of the present result will be provided hereafter. The result is hoped to be theoretically justified via extending the line of reasoning used in [49, 136] for nonlinear singularly perturbed systems to a class of discontinuous systems; its rigorous proof is left to the interested reader as an open problem.

8.5 Experimental Study

8.5.1 Experimental Setup

The experimental setup, designed in the research laboratory of CITEDI-IPN and shown in Fig. 8.1, involves a 3-DOF industrial robot manipulator manufactured by Amatrol. Figure 8.2 describes the entire hardware setup configuration of the system. The base of the mechanical robot has a horizontal revolute joint q_1, whereas two links have vertical revolute joints q_2 and q_3. The parameters for the mechanical manipulator are summarized in Table 8.1. The worm gear set, helicon gear set, and roller chain are used for torque transmission to joints q_1, q_2, and q_3, respectively; there is a DC gear motor for each joint, with a reduction ratio of 19.7:1 for q_1 and q_2, and 127.8:1 for q_3. These gears are the main source of friction. The ISA Bus servo I/O card from the *Servo To Go* company is used for the real-time control system. It consists of 8 channels of 16-bit D/A output, 32 bits of digital I/O, and an interval timer capable of interrupting the PC. The controller is implemented using the C++ programming language running on a 486 PC. Position measurements of each articulation of the robot are obtained using the channels of quadrature encoders

8.5 Experimental Study

Fig. 8.1 The 3-DOF robot manipulator

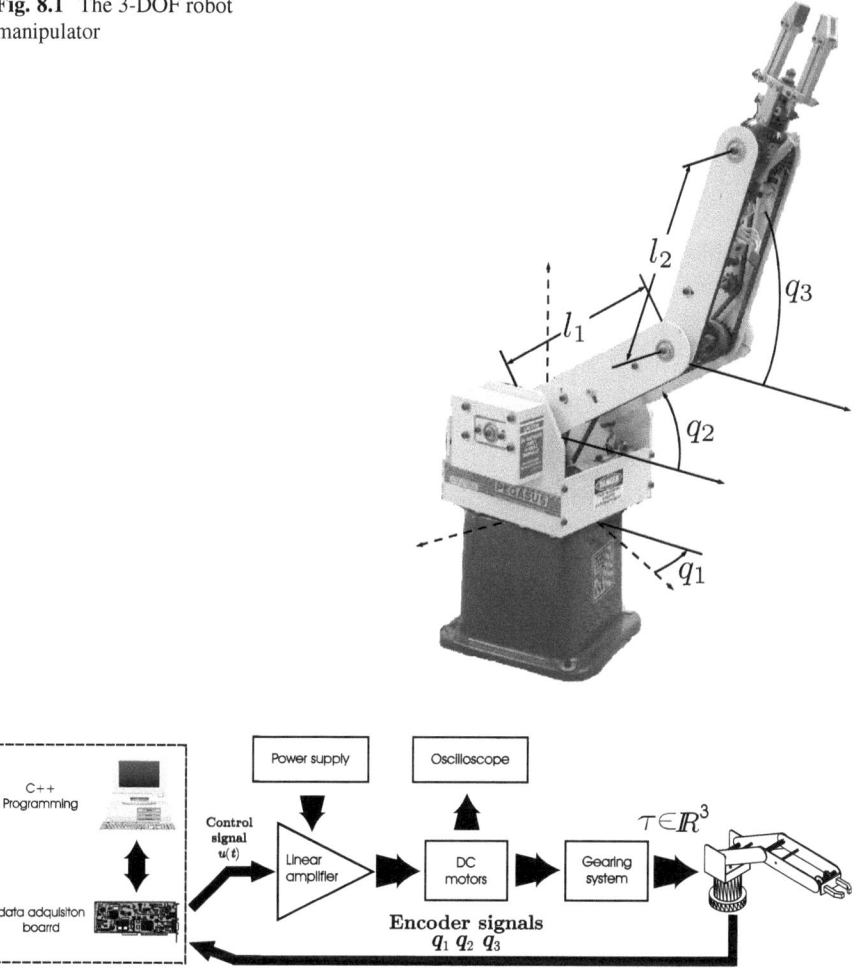

Fig. 8.2 Experimental setup

available on each DC gear motor, which is connected to the I/O card, programmed to provide the encoder signal processing each millisecond; the resolution of each encoder is 52×10^{-3} rad, 62×10^{-3} rad, and 34×10^{-3} rad for q_1, q_2, and q_3, respectively. Along with this, a digital oscilloscope is utilized to store control signals. Linear power amplifiers are installed in each servomotor, which applies a variable torque to each joint. These amplifiers accept control inputs from a D/A converter in the range of ± 10 V.

8.5.2 Three-DOF Manipulator Model

The motion of the experimental manipulator, governed by (8.10), was specified by applying the Euler–Lagrange formulation, where

$$M(q) = \begin{bmatrix} M_{11} & 0 & 0 \\ 0 & M_{22} & M_{23} \\ 0 & M_{23} & M_{33} \end{bmatrix}$$

$$M_{11}(q) = m_2 l_1^2 \cos^2(q_2) + 2m_2 l_1 l_2 \cos(q_2) \cos(q_2 + q_3)$$
$$\qquad\qquad + m_2 l_2^2 \cos^2(q_2 + q_3) + m_1 l_1^2 \cos^2(q_2) + I_1,$$
$$M_{22}(q) = m_1 l_1^2 + m_2 l_1^2 + 2m_2 l_1 l_2 \cos(q_3) + m_2 l_2^2$$
$$\qquad\qquad + I_2 + I_3,$$
$$M_{23}(q) = m_2 l_1 l_2 \cos(q_3) + m_2 l_2^2 + I_3,$$
$$M_{33}(q) = m_2 l_2^2 + I_3;$$

$$C(q, \dot{q}) = \begin{bmatrix} C_{11} & C_{12} & C_{13} \\ C_{21} & C_{22} & C_{23} \\ C_{31} & C_{32} & 0 \end{bmatrix}$$

$$C_{11} = -m_2 l_1^2 S_2 C_2 \dot{q}_2 - m_2 l_1 l_2 S_2 C_{23} \dot{q}_2 - m_1 l_1^2 S_2 C_2 \dot{q}_2$$
$$\qquad - m_2 l_1 l_2 C_2 S_{23}(\dot{q}_2 + \dot{q}_3) - m_2 l_2^2 S_{23} C_{23}(\dot{q}_2 + \dot{q}_3)$$
$$C_{12} = -m_2 l_1^2 S_2 C_2 \dot{q}_1 - m_2 l_1 l_2 C_2 S_{23} \dot{q}_1 - m_2 l_1 l_2 S_2 C_{23} \dot{q}_1$$
$$\qquad - m_2 l_2^2 S_{23} C_{23} \dot{q}_1 - m_1 l_1^2 S_2 C_2 \dot{q}_1,$$
$$C_{13} = -m_2 l_1 l_2 C_2 S_{23} \dot{q}_1 - m_2 l_2^2 S_{23} C_{23} \dot{q}_1,$$
$$C_{21} = m_2 l_1^2 S_2 C_2 \dot{q}_1 + m_2 l_1 l_2 C_2 S_{23} \dot{q}_1 + m_2 l_1 l_2 S_2 C_{23} \dot{q}_1$$
$$\qquad + m_2 l_2^2 S_{23} C_{23} \dot{q}_1 + m_1 l_1^2 S_2 C_2 \dot{q}_1$$
$$C_{22} = -m_2 l_1 l_2 S_3 \dot{q}_3,$$
$$C_{23} = -m_2 l_1 l_2 S_3 (\dot{q}_2 + \dot{q}_3),$$
$$C_{31} = m_2 l_1 l_2 C_2 S_{23} \dot{q}_1 + m_2 l_2^2 S_{23} C_{23} \dot{q}_1,$$
$$C_{32} = m_2 l_1 l_2 S_3 \dot{q}_2;$$

8.5 Experimental Study

Table 8.1 Parameters of the mechanical manipulator

Description	Notation	Value	Units
Length of link 1	l_1	0.297	m
Length of link 2	l_2	0.297	m
Mass of link 1	m_1	0.38	kg
Mass of link 2	m_2	0.34	kg
Inertia 1	I_1	0.243×10^{-3}	kg m^2
Inertia 2	I_2	0.068×10^{-3}	kg m^2
Inertia 3	I_3	0.015×10^{-3}	kg m^2
Gravity acceleration	g	9.8	m/seg^2

with $S_i = \sin q_i$, $C_i = \cos q_i$, $S_{ij} = \sin(q_i + q_j)$, $C_{ij} = \cos(q_i + q_j)$;

$$G(q) = g \begin{bmatrix} 0 \\ m_1 l_1 \cos q_2 + m_2 l_1 \cos q_2 + m_2 l_2 \cos(q_2 + q_3) \\ m_2 l_2 \cos(q_2 + q_3) \end{bmatrix}.$$

The physical constant parameters, m_i, l_i, $i = 1, 2$, and I_j, $j = 1, 2, 3$, are given in Table 8.1.

8.5.3 Experimental Results

The regulator performance was studied experimentally. For comparison, four $\mathcal{H}_\infty$ controllers were implemented, namely,

- a nonsmooth controller (8.29)–(8.31), the design of which was based on the Dahl friction model,
- a discontinuous controller (8.41)–(8.43), the design of which was based on the Coulomb friction model,
- a nonlinear controller (8.41)–(8.43) subject to $F_c = 0$, the design of which was based on the viscous friction model without counting for any dry friction force,
- a linear controller (1.49), specified with (4.37), (8.41), (8.42) under $F_c = 0$, which was designed for the linearized manipulator model, counting for viscous friction only.

The viscous friction coefficient σ_0, stiffness σ_1, and Coulomb friction level F_c were obtained by applying the procedure from [70]:

$$\sigma_0 = diag\{9.84, 13.02, 9.87\} \text{Nms/rad},$$
$$\sigma_1 = diag\{0.054, 0.053, 0.039\} \text{Nm},$$
$$F_c = diag\{2.1, 1.02, 0.78\} \text{Nm}.$$

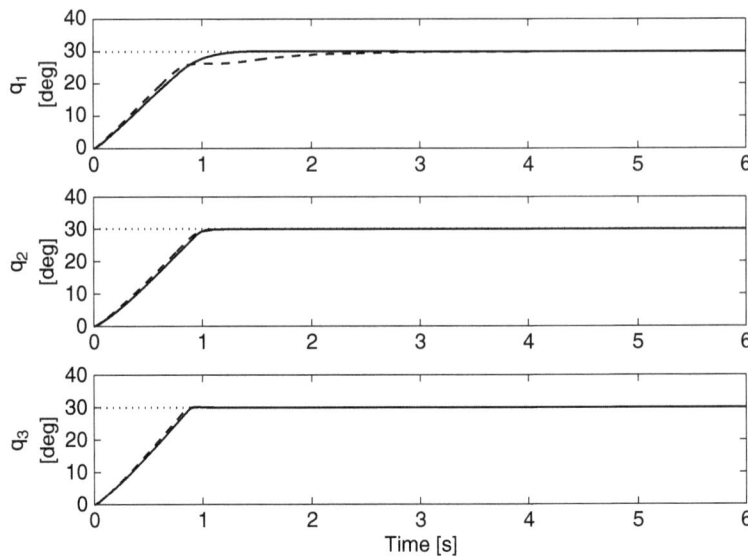

Fig. 8.3 Experimental results for the nonsmooth (*solid lines*) and linear (*dotted lines*) $\mathcal{H}_\infty$ regulators

First, the robot manipulator was required to move in the space from the origin $q_1 = q_2 = q_3 = 0$ (the references for each joint are shown in Fig. 8.1) to the desired position $q_{d1} = q_{d2} = q_{d3} = 30°$. The initial velocity $\dot{q}(0)$ and compensator state $\xi(0)$ were set to zero for all the experiments.

We achieved the control goal by implementing the nonsmooth $\mathcal{H}_\infty$ regulator with the weight parameter $\rho = 10$ on the 3-DOF manipulator. By iterating on γ, we found the infimal achievable level $\gamma^* \simeq 9$. However, in the subsequent experiment, we selected $\gamma = 125$ to avoid an undesirable high-gain controller design that would appear for a value of γ close to the optimum. With $\gamma = 125$, we obtained that for $\varepsilon = 0.1$, the corresponding Riccati equations (6.64), (6.65) have positive definite solutions. By Theorem 34, these solutions result in the control law (8.35) solving the regulation problem. For the sake of comparison, the linear and nonlinear $\mathcal{H}_\infty$ regulators, synthesized for the viscous friction-based model of the robot, have also been tested on the laboratory manipulator. The resulting trajectories are depicted in Figs. 8.3 and 8.4, respectively. These figures demonstrate that the nonsmooth $\mathcal{H}_\infty$ regulator does asymptotically stabilize the system motion around the desired position, as opposed to the standard linear and nonlinear $\mathcal{H}_\infty$ regulators that drive the robot to wrong endpoints. Thus, we may conclude that the nonsmooth $\mathcal{H}_\infty$ regulator has a better performance.

In addition, we successively applied the nonsmooth $\mathcal{H}_\infty$ regulator to the robot manipulator with 5-Nm overload in each joint (thereby modeling permanent external disturbances), and to the manipulator whose nominal parameter values (length, mass, Coulomb level, stiffness, and viscous friction) had successively been varied in

8.5 Experimental Study

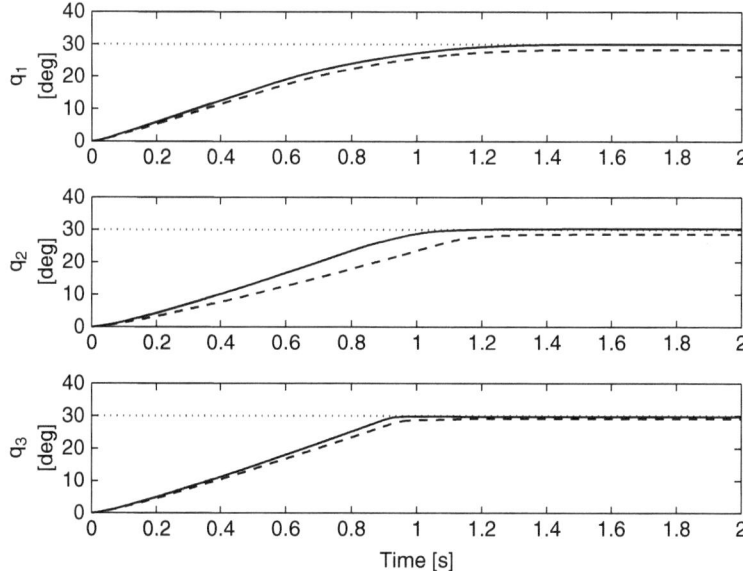

Fig. 8.4 Experimental results for the nonsmooth (*solid lines*) and nonlinear (*dotted lines*) $\mathcal{H}_\infty$ regulators

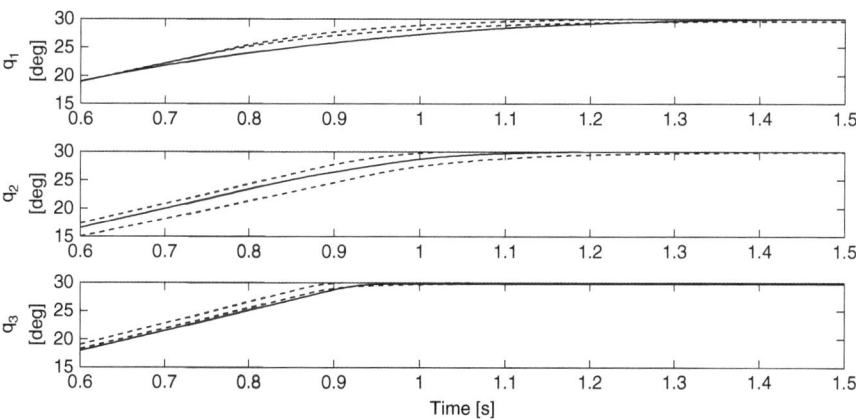

Fig. 8.5 Robustness of the nonsmooth $\mathcal{H}_\infty$ regulator against parametric uncertainties: *Solid line* is for the nominal parameter values; *dotted lines* for $\pm 30\%$ parameter variations

the 30 % and −30 % ranges. The resulting motion of the closed-loop system with the overload and that with the modified parameters are shown in Fig. 8.5. These figures carry out favorable robustness properties of the closed-loop system, driven by the nonsmooth regulator against external disturbances and parameter variations.

Next, a discontinuous $\mathcal{H}_\infty$ regulator was implemented. Since this regulator counted for static viscous and Coulomb friction forces only, it was of a lower

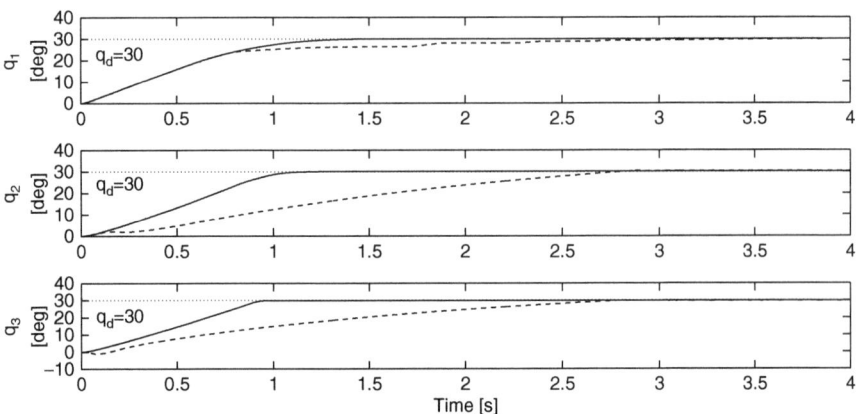

Fig. 8.6 Experimental results for the nonsmooth $\mathcal{H}_\infty$ (*solid lines*) regulator and its discontinuous version (*dotted lines*)

order compared to the nonsmooth controller, and therefore it allowed an easier implementation. Figure 8.6 shows that the discontinuous regulator brought the robot manipulator to the desired position ($q_{di} = 30°$ ($i = 1, 2, 3$)) as well; however, it was done at the price of slower dynamics compared to the nonsmooth regulator.

In order to speed up the stabilization process and address the regulation problem for initial conditions, which did not fall into the attraction domain of a desired endpoint, we additionally implemented nonsmooth and discontinuous $\mathcal{H}_\infty$-tracking controllers, selecting the weight parameter $\rho = 10$ and specifying the desired trajectory to follow as

$$q_{di}(t) = \mu_i(1 - e^{-\omega_i t}), \quad i = 1, 2, 3. \tag{8.50}$$

Dynamic properties of the above trajectory were prescribed as $\omega_i = 1.8$ ($i = 1, 2, 3$) to run reasonably fast initially toward the desired endpoint $q_d^T = [\mu_1, \mu_2, \mu_3]$ and quietly enough near this endpoint. The desired trajectory was initialized in the origin $q_{di}(0) = 0$ ($i = 1, 2, 3$), thereby yielding the initial state deviation $x_{1i}(0) = q_i(0) - q_{di}(0) = 0$ independent of μ_i. The desired endpoint $\mu_i = 90$, $i = 1, 2, 3°$ was prespecified far from the origin. The initial velocity $\dot{q}(0)$, Dahl friction force $F_d(0)$, and compensator state $\xi(0)$ were equivalent to zero in the experiment.

Since we chose the to-be-tracked trajectory, leading to the desired endpoint, to be smooth with uniformly bounded (in t) derivatives, Theorems 33 and 35 were applicable to the corresponding $\mathcal{H}_\infty$-tracking problem, and the nonsmooth controller (8.31) and the discontinuous controller (8.43), once properly specified, solved the problem. By iterating on γ, we found the infimal achievable attenuation level γ^* for the nonsmooth tracking controller and then for the discontinuous one. These values were approximately $\gamma^* \simeq 10$ for both controllers. However, in the subsequent simulation, we selected $\gamma = 125$ to avoid an undesirable high-gain

8.5 Experimental Study

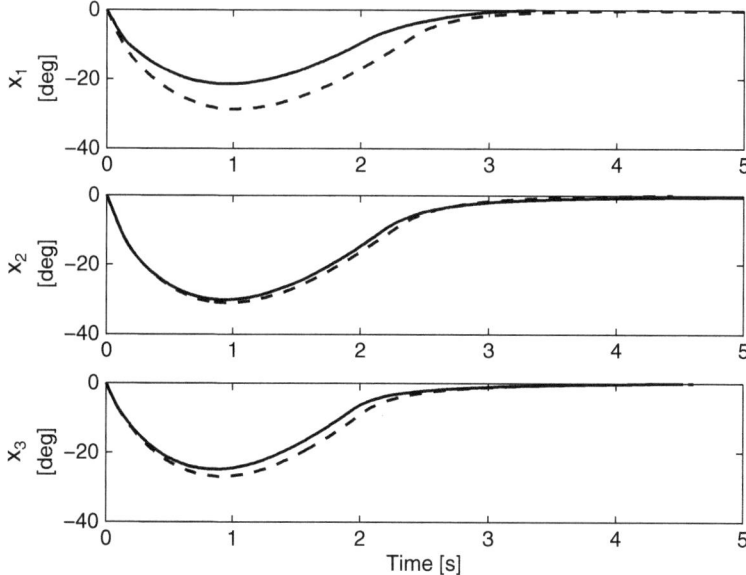

Fig. 8.7 Experimental results for the nonsmooth (*solid lines*) and discontinuous (*dotted lines*) $\mathcal{H}_\infty$-tracking controllers

controller design that would appear for values of γ close to the optimum. We obtained that under $\gamma = 125$, the Riccati equations, corresponding to the nonsmooth and discontinuous controllers' design, possessed uniformly bounded (in t) positive definite solutions for $\varepsilon = 1$. By Theorems 33 and 35, the corresponding solutions resulted in the nonsmooth (8.31) and discontinuous (8.43) control laws, solving the tracking problem. A good performance, concluded from Fig. 8.7, was obtained for both nonsmooth and discontinuous controllers. As in the regulation case, the nonsmooth controller yielded a faster convergence to the desired endpoint, whereas the discontinuous controller admitted an easier implementation.

Finally, the periodic tracking was tested for the following trajectory to follow:

$$q_{di} = 25\sin(\omega t), \qquad i = 1, 2, 3. \tag{8.51}$$

The same parameter values, $\rho = 10$, $\gamma = 125$, $\varepsilon = 1$, were appropriate for the nonsmooth and discontinuous controller designs. The experimental results, depicted in Figs. 8.8 and 8.9, illustrated capabilities of the proposed nonsmooth and discontinuous designs for adequate periodic tracking.

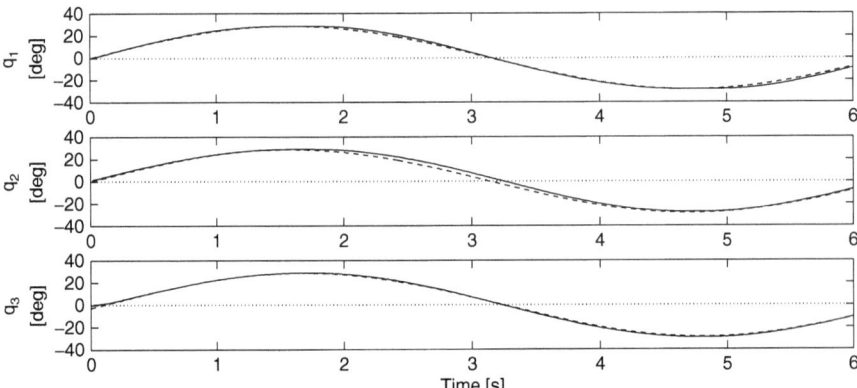

Fig. 8.8 Experimental results for the periodic nonsmooth $\mathcal{H}_\infty$-tracking controller: *Solid line* is for the periodic trajectory to be tracked; *dotted line* is for the closed-loop output

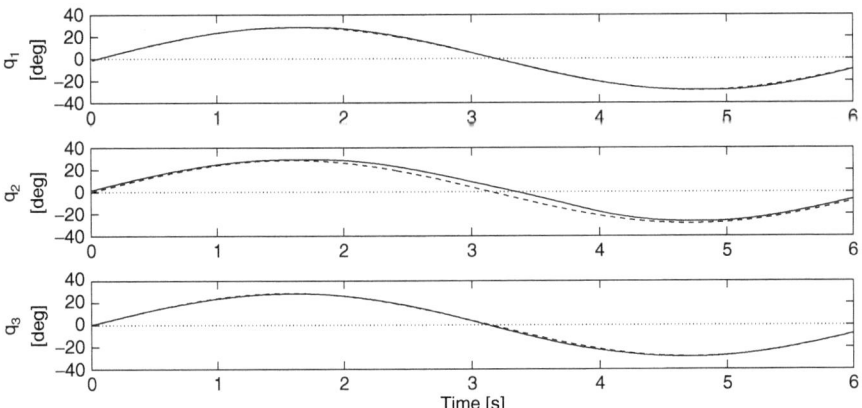

Fig. 8.9 Experimental results for the periodic discontinuous $\mathcal{H}_\infty$ tracking controller: *Solid line* is for the periodic trajectory to be tracked; *dotted line* is for the closed-loop output

8.6 Synthesis over Nonlinear Sampled-Data Measurements: A Case Study

In order to illustrate the $\mathcal{H}_\infty$ synthesis over sampled-data measurements, we present a simulation study, conducted in [16], for an inverted pendulum, moving in the vertical plane and depicted in Fig. 8.10.

The pendulum is driven by an actuator u, using position measurements only. The measurements, which are available at the discrete-time moments $\tau_j = 0.5j$, $j = 0, 1, \ldots$, are made by a nonlinear potentiometer with the standard resistance function $\sin \theta$. The aim is to design a $\mathcal{H}_\infty$ sampled-data measurement feedback $\mathcal{H}_\infty$ controller, locally stabilizing the inverted pendulum around its unstable upright

8.6 Synthesis over Nonlinear Sampled-Data Measurements: A Case Study

Fig. 8.10 Inverted pendulum

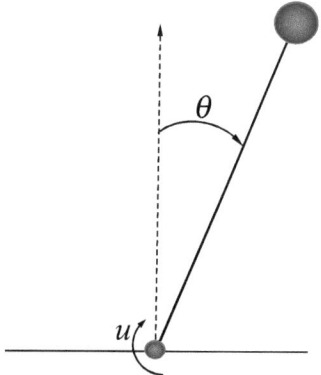

equilibrium, whose dimensionless model is given in terms of the angular position $x_1 = \theta$ and angular velocity $x_2 = \dot{\theta}$:

$$\dot{x}_1 = x_2,$$
$$\dot{x}_2 = \sin(x_1) + w_1 + u,$$
$$z_1 = u, \; z_2 = x_1,$$
$$y_j = \sin x_1(\tau_j) + w_2(\tau_j), \quad j = 0, 1, \ldots. \tag{8.52}$$

According to Theorem 27, constituting the $\mathcal{H}_\infty$ synthesis of Sect. 6.3 over sampled-data measurements, a local solution of the sampled-data measurement $\mathcal{H}_\infty$-control problem for (8.52) is constructed as follows:

$$\begin{bmatrix} \dot{\xi}_1 \\ \dot{\xi}_2 \end{bmatrix} = \begin{bmatrix} \xi_2 \\ \sin(\xi_1) \end{bmatrix} + \left\{ \frac{1}{\gamma^2} \begin{bmatrix} 0 & 0 \\ 0 & 1 \end{bmatrix} - \begin{bmatrix} 0 & 0 \\ 0 & 1 \end{bmatrix} \right\} P_\varepsilon \begin{bmatrix} \xi_1 \\ \xi_2 \end{bmatrix},$$

$$\xi_i(\tau_j+) = \varsigma_{ij}(1), \quad i = 1, 2, \quad j = 0, 1, \ldots,$$

$$u = -\begin{bmatrix} 0 & 1 \end{bmatrix} P_\varepsilon \begin{bmatrix} \xi_1 \\ \xi_2 \end{bmatrix}, \tag{8.53}$$

where ς_{ij}, P_ε, and Z_ε satisfy the equations

$$\begin{bmatrix} \dot{\varsigma}_{1j} \\ \dot{\varsigma}_{2j} \end{bmatrix} = Z_\varepsilon(\tau_j-) \left\{ \begin{bmatrix} 1 & 0 \\ 0 & 1 \end{bmatrix} + \begin{bmatrix} 1 & 0 \\ 0 & 0 \end{bmatrix} Z_\varepsilon(\tau_j-)t \right\}^{-1} \times$$

$$\begin{bmatrix} 1 \\ 0 \end{bmatrix} (y(\tau_j) - \sin(v_1)); \quad \begin{bmatrix} \varsigma_{1j}(0) \\ \varsigma_{2j}(0) \end{bmatrix} = \begin{bmatrix} \xi_1(\tau_j-) \\ \xi_2(\tau_j-) \end{bmatrix}, \quad j = 0, 1, \ldots, \tag{8.54}$$

$$P_\varepsilon \begin{bmatrix} 0 & 1 \\ 1 & 0 \end{bmatrix} + \begin{bmatrix} 0 & 1 \\ 1 & 0 \end{bmatrix}^T P_\varepsilon + \begin{bmatrix} 1 & 0 \\ 0 & 0 \end{bmatrix}$$

$$+ P_\varepsilon \left\{ \frac{1}{\gamma^2} \begin{bmatrix} 0 & 0 \\ 0 & 1 \end{bmatrix} - \begin{bmatrix} 0 & 0 \\ 0 & 1 \end{bmatrix} \right\} P_\varepsilon + \varepsilon \begin{bmatrix} 1 & 0 \\ 0 & 1 \end{bmatrix} = 0, \quad (8.55)$$

$$\dot{Z}_\varepsilon = \tilde{A}_\varepsilon Z_\varepsilon(t) + Z_\varepsilon(t) \tilde{A}_\varepsilon^T + \begin{bmatrix} 0 & 0 \\ 0 & 1 \end{bmatrix}$$

$$+ \frac{1}{\gamma^2} Z_\varepsilon(t) P_\varepsilon \begin{bmatrix} 0 & 0 \\ 0 & 1 \end{bmatrix} P_\varepsilon Z_\varepsilon(t) + \varepsilon \begin{bmatrix} 1 & 0 \\ 0 & 1 \end{bmatrix},$$

$$Z_\varepsilon(\tau_j+) = Z_\varepsilon(\tau_j-) \left\{ \begin{bmatrix} 1 & 0 \\ 0 & 1 \end{bmatrix} + \begin{bmatrix} 1 & 0 \\ 0 & 0 \end{bmatrix} Z_\varepsilon(\tau_j-) \right\}^{-1}, \quad (8.56)$$

with

$$\tilde{A}_\varepsilon = \begin{bmatrix} 0 & 1 \\ 1 & 0 \end{bmatrix} + \frac{1}{\gamma^2} \begin{bmatrix} 0 & 0 \\ 0 & 1 \end{bmatrix} P_\varepsilon$$

and γ provided that Eqs. (8.55), (8.56), corresponding to $\varepsilon = 0$, have uniformly bounded positive semidefinite symmetric solutions such that the system

$$\dot{X}_P = \left\{ \begin{bmatrix} 0 & 1 \\ 1 & 0 \end{bmatrix} - \left(\begin{bmatrix} 0 & 0 \\ 0 & 1 \end{bmatrix} - \frac{1}{\gamma^2} \begin{bmatrix} 0 & 0 \\ 0 & 1 \end{bmatrix} \right) P \right\} \dot{X}_P,$$

$$\dot{X}_Z = \left\{ \tilde{A} + \frac{1}{\gamma^2} ZP \begin{bmatrix} 0 & 0 \\ 0 & 1 \end{bmatrix} P \right\} X_Z,$$

$$X_Z(\tau_j+) = X_Z(\tau_j-) - Z(\tau_j-) \left\{ \begin{bmatrix} 1 & 0 \\ 0 & 1 \end{bmatrix} + \begin{bmatrix} 1 & 0 \\ 0 & 0 \end{bmatrix} Z(\tau_j-) \right\}^{-1} \begin{bmatrix} 1 & 0 \\ 0 & 0 \end{bmatrix} X_Z(\tau_j-)$$

is exponentially stable.

In the simulation, performed with MATLAB®, using the $\mathcal{H}_\infty$-design procedure of Sect. 6.3 yields $\gamma^* \approx 5.32$. However, $\gamma = 10$ is utilized in the simulation to avoid an undesired high-gain controller design corresponding to a value of γ close to the optimum. For the Riccati equation (8.55), corresponding to this value of γ and $\varepsilon = 0.1$, the positive definite symmetric solution

$$P_\varepsilon = \begin{bmatrix} 3.261145 & 2.470038 \\ 2.470038 & 2.256321 \end{bmatrix}$$

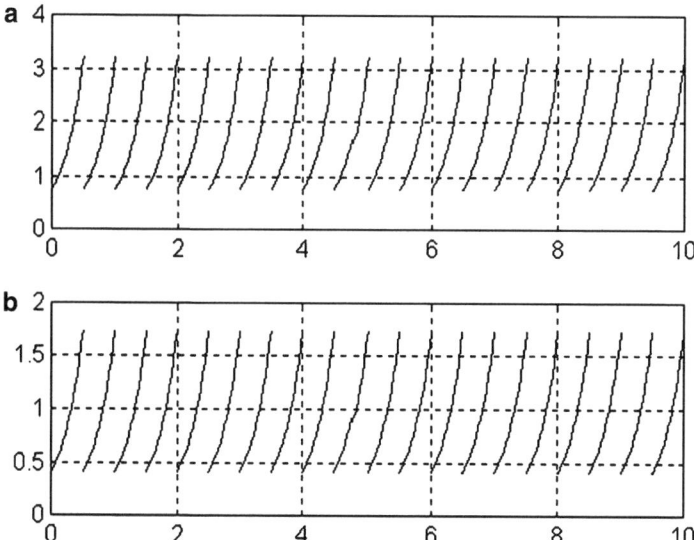

Fig. 8.11 Plot of minors of $Z_\varepsilon(t)$: (**a**) the first-order minor and (**b**) the second-order minor

is computed, and the uniformly bounded positive definite symmetric solution $Z_\varepsilon(t)$, the minors of which are presented in Fig. 8.11, is then derived from (8.56).

These solutions result in the control law (8.53), which solves the problem in question. Two cases for the controller, with no disturbances and with permanent disturbances $w_1 = 0.08$ and $w_2 = 0.07$, are simulated. The initial conditions for the simulations are set to $x_1(0) = 20°$, $x_2(0) = 0$, $\xi_1(0) = \xi_2(0) = 0$. The resulting control inputs and trajectories are depicted in Figs. 8.12 and 8.13.

From Fig. 8.12, we conclude that in spite of the large position deviation $x_1(0) = 20°$ from the desired one, the pendulum moves reasonably fast initially toward its endpoint; however, the final positioning requires more time. Figure 8.13 shows that a good system performance is still provided, while the permanent disturbances enforce the system.

8.7 Concluding Remarks

The nonsmooth $\mathcal{H}_\infty$ synthesis of Sect. 6.1 was first developed for robot manipulators with frictional joints and it was then extended to the discontinuous synthesis for counting for dry friction forces. The experimental study, made for an industrial 3-DOF robot manipulator, clearly justified the superiority of the proposed nonsmooth and discontinuous $\mathcal{H}_\infty$ designs versus the standard linear and nonlinear $\mathcal{H}_\infty$ designs. The former designs counted for the dynamic Dahl model and static Coulomb model of dry friction forces, whereas in the latter designs, the dry

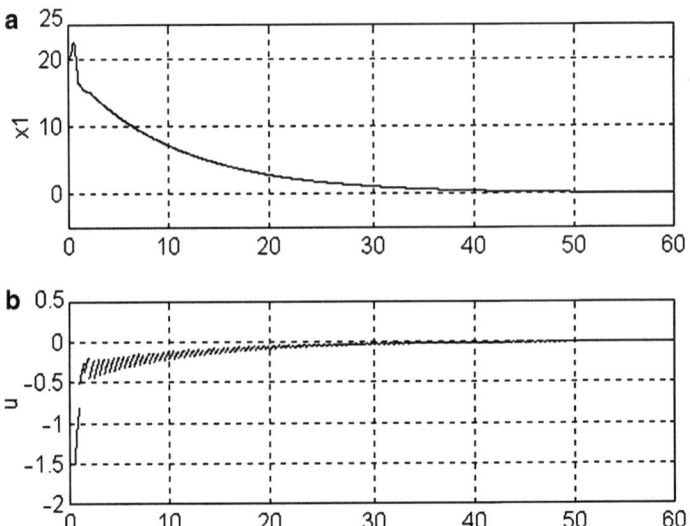

Fig. 8.12 The no-disturbance case: plots of (**a**) the angular position and (**b**) the control input

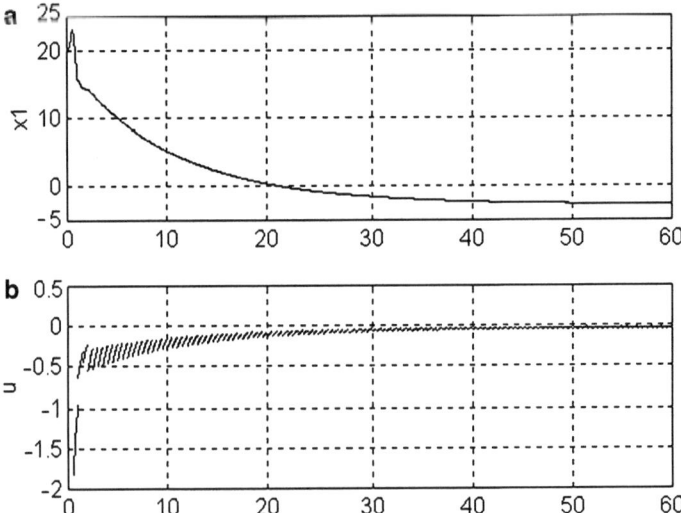

Fig. 8.13 The permanent disturbance case: plots of (**a**) the angular position and (**b**) the control input

friction forces were ignored. The discontinuous synthesis was shown to be easier in implementation, whereas the nonsmooth synthesis was capable of resulting in a faster convergence to a desired endpoint and a higher precision in periodic tracking.

8.7 Concluding Remarks

In addition, the effectiveness of the sampled-data $\mathcal{H}_\infty$ synthesis of Sect. 6.3 was illustrated by stabilizing an inverted pendulum around its unstable upright equilibrium in the situation where position measurements, made by a nonlinear potentiometer, were available at equidistant discrete-time instants only.

Chapter 9
Nonsmooth $\mathcal{H}_\infty$ Synthesis in the Presence of Backlash

A number of industrial systems and consumer products currently in use are characterized by the actuator nonidealities, such as backlash, dead zone, and others, adversely affecting system performance. Consistently addressing these nonsmooth nonidealities, while also providing robust performance, has been a longstanding problem in the control of these plants. This problem is presently addressed through the nonsmooth $\mathcal{H}_\infty$-output controller design of Chap. 6, which takes into account the dead zone in the transmitted actuator signal. The nonsmooth $\mathcal{H}_\infty$-output regulator is first developed for a servomechanism, driven by a DC motor, where backlash effects occur in the servomechanism itself, and then it is applied to the experimentally obtained boiler/turbine model with an actuator dead zone. The simulation results showing a considerable performance improvement are given.

9.1 Output Regulation of a Servomechanism with Backlash

In the present section, the output regulation problem is studied for an electrical actuator consisting of a motor part driven by a DC motor and a reducer part (load) operating under uncertainty conditions in the presence of nonsmooth backlash effects, resulting in the multistability of the system in the open loop. The objective is to drive the load to a desired position while providing the boundedness of the system motion and attenuating external disturbances. Due to practical requirements [76], the motor angular position is assumed to be the only information available for feedback.

The above problem is locally resolved within the nonsmooth $\mathcal{H}_\infty$-control approach of Chap. 6 as opposed to [3, 4], where a smooth single-stability approximation of the backlash model is utilized (thereby severely limiting the achievable performance) and the standard nonlinear $\mathcal{H}_\infty$ approach is then applied. It is worth noticing that both the developed approach and the standard one do not admit a straightforward application to the above problem because a partial state stabilization (i.e., asymptotic stabilization of the output of the system) is required only provided that the complementary variables remain bounded. The fact that under motor position measurements the servomotor with backlash presents a nonminimum phase system is another challenge to be overcome. The better performance of the proposed nonsmooth $\mathcal{H}_\infty$ controller compared to the standard one, developed for the smoothed backlash model, is illustrated in an experimental study made for a DC motor linked to a mechanical load through an imperfect contact gear train.

9.1.1 Dead Zone–Based Model of Backlash Phenomenon

Backlash occurs in any mechanical system where a driving part (motor) is not directly connected with a driven part (load). An electrical actuator consists of a DC motor and a load coupled by a gear reduction part. The dynamics of the angular position $q_i(t)$ of the DC motor and of the load $q_o(t)$ are modeled as follows [81]:

$$J_o N^{-1} \ddot{q}_o + f_o N^{-1} \dot{q}_o = T + w_o,$$

$$J_i \ddot{q}_i + f_i \dot{q}_i + T = \tau_m + w_i. \tag{9.1}$$

Hereinafter, J_o, f_o, $\ddot{q}_o$, and $\dot{q}_o$ are, respectively, the inertia of the load and the reducer, the viscous output friction, the output acceleration, and the output velocity. The inertia of the motor, the viscous motor friction, the motor acceleration, and the motor velocity are denoted by J_i, f_i, $\ddot{q}_i$, and $\dot{q}_i$, respectively. The input torque τ_m serves as a control action, and T stands for the transmitted torque. The external disturbances $w_i(t)$, $w_o(t)$ have been introduced into the driver equation (9.1) to account for destabilizing model discrepancies due to hard-to-model nonlinear backlash phenomena. While controlling the servomotor, the influence of the disturbances $w_i(t)$, $w_o(t)$ on the performance of the closed-loop system is to be attenuated.

The transmitted torque T through a backlash with an amplitude j is typically modeled by a dead zone characteristic [87, p. 8]:

$$T(\Delta q) = \begin{cases} 0 & \text{if } |\Delta q| \le j, \\ K \Delta q - K j \, sign(\Delta q) & \text{otherwise,} \end{cases} \tag{9.2}$$

where

$$\Delta q = q_i - N q_o, \tag{9.3}$$

K is the stiffness, and N is the reducer ratio. Such a model is depicted in Fig. 9.1. Provided the servomotor position $q_i(t)$ is the only available measurement on the system, the above model (9.1)–(9.3) is of nonminimum phase because, along with the origin, the unforced system possesses a multivalued set of equilibria (q_i, q_o) with $q_i = 0$ and $q_o \in [-j, j]$.

9.1 Output Regulation of a Servomechanism with Backlash

Fig. 9.1 The dead zone model of backlash

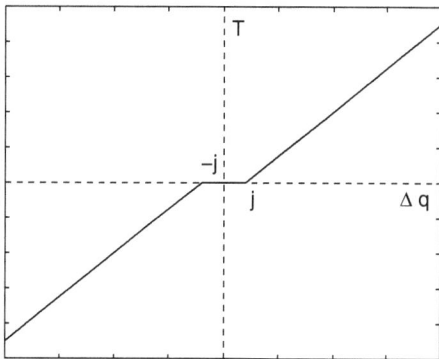

9.1.2 Problem Statement and Control Synthesis

The objective of the $\mathcal{H}_\infty$-output regulation of the nonlinear drive system (9.1) with backlash effects (9.2)–(9.3) is to design a nonlinear $\mathcal{H}_\infty$ controller so as to obtain the closed-loop system in which all these trajectories are bounded and the output $q_o(t)$ asymptotically decays to a desired position q_d as $t \to \infty$ while also attenuating the influence of the external disturbances $w_i(t), w_o(t)$.

To formally state the problem, we'll introduce the state deviation vector $x = [x_1, x_2, x_3, x_4]^T$ with

$$x_1 = q_o - q_d, \quad x_2 = \dot{q}_o, \quad x_3 = q_i - Nq_d, \quad x_4 = \dot{q}_i,$$

where x_1 is the load position error, x_2 is the load velocity, x_3 is the motor position deviation from its nominal value, and x_4 is the motor velocity. The nominal motor position Nq_d has been prespecified in such a way to guarantee that $\Delta q = \Delta x$, where

$$\Delta x = x_3 - N x_1.$$

Then system (9.1)–(9.3), represented in terms of the deviation vector x, takes the form

$$\begin{aligned}
\dot{x}_1 &= x_2, \\
\dot{x}_2 &= J_o^{-1}[NT(\Delta x) - f_o x_2 + N w_o], \\
\dot{x}_3 &= x_4, \\
\dot{x}_4 &= J_i^{-1}[\tau_m - T(\Delta x) - f_i x_4 + N w_i].
\end{aligned} \quad (9.4)$$

The zero dynamics

$$\dot{x}_1 = x_2,$$
$$\dot{x}_2 = J_o^{-1}[NT(-Nx_1) - f_o x_2] \quad (9.5)$$

of the undisturbed version of system (9.4) with respect to the output

$$y = x_3 \quad (9.6)$$

is formally obtained by specifying the control law that maintains the output identically zero.

Taking into account that the transmitted torque (9.2) is governed by a nonsmooth function, the $\mathcal{H}_\infty$-output regulation problem for the drive system with backlash can formally be stated as a nonsmooth $\mathcal{H}_\infty$-control problem for the deviation system (9.4). In the sequel, we confine our investigation to the $\mathcal{H}_\infty$-position regulation problem, where

1. The output to be controlled is given by

$$z = \rho \begin{bmatrix} 0 \\ x_1 \end{bmatrix} + \begin{bmatrix} 1 \\ 0 \end{bmatrix} \tau_m, \quad (9.7)$$

with a positive weight coefficient ρ.

2. The motor position q_i is the only available measurement and this measurement is corrupted by the error vector $w_y(t) \in \mathbb{R}$; that is,

$$y = x_3 + Nq_d + w_y. \quad (9.8)$$

The $\mathcal{H}_\infty$-control problem in question is thus stated as follows. Given the system representation (9.4)–(9.8) and a real number $\gamma > 0$, one is required to find (if any) a causal dynamic feedback controller

$$u = \mathcal{K}(\xi) \quad (9.9)$$

with internal state $\xi \in \mathbb{R}^4$ such that the undisturbed closed-loop system is uniformly asymptotically stable around the origin and its $\mathcal{L}_2$ gain is locally less than γ; namely, the inequality

$$\int_0^T \|z(t)\|^2 dt < \gamma^2 \int_0^T \|w(t)\|^2 dt \quad (9.10)$$

holds for all $T > 0$ and all piecewise-continuous functions $w^T(t) = [w_o, w_i, w_y](t)$ for which the corresponding state trajectory of the closed-loop system, initialized at the origin, remains in some neighborhood of this point.

9.1 Output Regulation of a Servomechanism with Backlash

If we set

$$f(x) = \begin{bmatrix} x_2 \\ J_o^{-1}(NT(x_3 - Nx_1) - f_o x_2) \\ x_4 \\ -J_i^{-1}(T(x_3 - Nx_1) + f_i x_4) \end{bmatrix}, \quad (9.11)$$

$$g_1(x) = \begin{bmatrix} 0 & 0 & 0 \\ NJ_o^{-1} & 0 & 0 \\ 0 & 0 & 0 \\ 0 & J_i^{-1} & 0 \end{bmatrix}, \qquad g_2(x) = \begin{bmatrix} 0_{3\times 1} \\ J_i^{-1} \end{bmatrix},$$

$$h_1(x) = \rho \begin{bmatrix} 0 \\ x_1 \end{bmatrix}, \qquad h_2(x) = x_3 + Nq_d,$$

$$k_{12}(x) = \begin{bmatrix} 1 \\ 0 \end{bmatrix}, \qquad k_{21}(x) = \begin{bmatrix} 0 & 0 & 1 \end{bmatrix}, \quad (9.12)$$

a local $\mathcal{H}_\infty$-position regulator of the servomechanism with backlash is synthesized along the line of the nonsmooth autonomous $\mathcal{H}_\infty$ synthesis of Sect. 6.2 by applying Theorem 26 to system (6.1)–(6.3) specified with (9.11), (9.12).

Theorem 36. *Let Conditions C1″ and C2″ of Sect. 6.2 hold for the matrix functions A, B_1, B_2, C_1, C_2, governed by (6.35), (9.11)–(9.12), and let $(P_\varepsilon, Z_\varepsilon)$ be the corresponding positive definite solution of (6.64), (6.65) under some $\varepsilon > 0$. Then the output feedback*

$$\dot{\xi} = f(\xi) + \left[\frac{1}{\gamma^2} g_1(\xi) g_1^T(\xi) - g_2(\xi) g_2^T(\xi) \right] P_\varepsilon \xi$$

$$+ Z_\varepsilon C_2^T [y - h_2(\xi)], \quad (9.13)$$

$$u = -g_2^T(\xi) P_\varepsilon \xi \quad (9.14)$$

specified with (9.11), (9.12) presents a local solution of the $\mathcal{H}_\infty$-position-regulation problem (9.4), (9.7)–(9.10).

Proof. Since the transmitted torque (9.2) is locally differentiable around the origin, and its derivative is nullified in the origin, all the assumptions of Sect. 6.1, including Assumption A5 of Sect. 6.1.5, are trivially verified for the state-error system (9.4), (9.7), (9.8), and Theorem 26 proves to be applicable to this system. By applying Theorem 26 to (9.4), (9.7), (9.8), we have then straightforwardly established the validity of the present theorem.

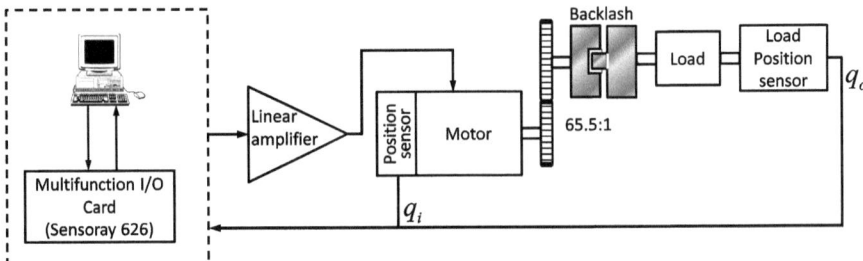

Fig. 9.2 Experimental test bench

Fig. 9.3 Backlash hysteresis before compensation

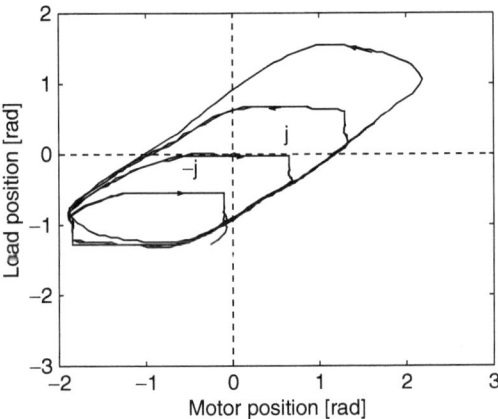

9.1.3 Experimental Study

9.1.3.1 Experimental Setup

Experimental setup, installed in the Robotics & Control Laboratory of CITEDI-IPN, involves a DC motor linked to a mechanical load through an imperfect contact gear train. Figure 9.2 shows the location of the sensors, actuator, and load. The input–output motion graph of Fig. 9.3 reveals the gear backlash effect.

The maximum allowable torque is 1.24 N m. The PCI Multifunction I/O board from *Sensoray 626* was used for the real-time control system. It consists of four channels of 14-bit D/A outputs and six quadrature 24-bit encoders. A shaft-mounted encoder gets the motor's angular position and provides the encoder signal to the I/O card each millisecond (sampling time). The resolution of the encoder is 2,000 ppr. A high-resolution potentiometer was placed on the load side to support the results. A linear power amplifier was installed in the servomotor to accept control signals from the D/A converter in the range of ± 10 V. The sampling time was 10×10^{-3} s.

9.1 Output Regulation of a Servomechanism with Backlash

Table 9.1 Nominal parameters

Description	Notation	Value	Units
Motor inertia	J_i	2.8×10^{-6}	kg m^2
Load inertia	J_o	1.07	kg m^2
Motor viscous friction	f_i	7.6×10^{-7}	N m s/rad
Load viscous friction	f_o	1.73	N m s/rad

For experimental purposes, we added a coupler to increase the backlash level to $j = 0.2$ rad. The gear reduction ratio is 65.5:1; it is the main source of friction. The stiffness coefficient K is computed experimentally by differentiating the transmitted torque measurements in the state variable at the zone of the linear growth.

Table 9.1 presents the parameters of the motor, taken from the manufacturer's data specifications, and the nominal load parameters, identified in accordance with the experimental procedures proposed in [70].

9.1.3.2 Experimental Results

The experiments were carried out for the closed-loop system (9.11)–(9.14) with a position sensor located at the motor side, thus considering the angular motor position as the only information available for feedback. In the experiments, the load was required to move from the initial static position $q_o(0) = 0$ to the desired position $q_d = \pi/4$ rad. In order to illustrate the size of the attraction domain, we chose the initial load position to be reasonably far from the desired position, whereas the initial velocity $\dot{q}_o(0)$ of the load and that $\dot{q}_i(0)$ of the motor, as well as the initial compensator state $\xi(0)$, were set to zero for all experiments. Despite disturbances and uncertainties that are ever-present in the implemented closed-loop system (such as sensor noise and modeling error), we additionally applied an unknown but bounded torque perturbation governed by

$$w_i(t) = 4e^{-t}\cos(20t). \tag{9.15}$$

The resulting trajectories and applied torque are depicted in Figs. 9.4 and 9.5, respectively. These figures demonstrate that the regulator stabilizes the disturbance-free load motion around the desired position and attenuates external load disturbances. As predicted, Fig. 9.4 shows that the load position goes asymptotically to the desired position while the motor position remains bounded. In Fig. 9.5, the load position shows a 2% error around its reference position when the torque is perturbed by (9.15). It should be noted that the potentiometer induces noise in the load position's measured signal, as reflected in Figs. 9.4 and 9.5.

158   9 Nonsmooth $\mathcal{H}_\infty$ Synthesis in the Presence of Backlash

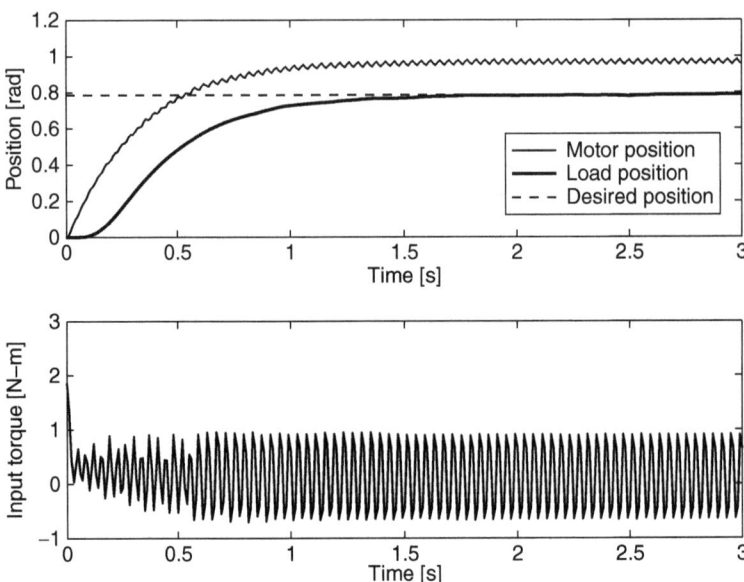

Fig. 9.4 Experimental results for the $\mathcal{H}_\infty$ regulator considering only the motor position information available for feedback

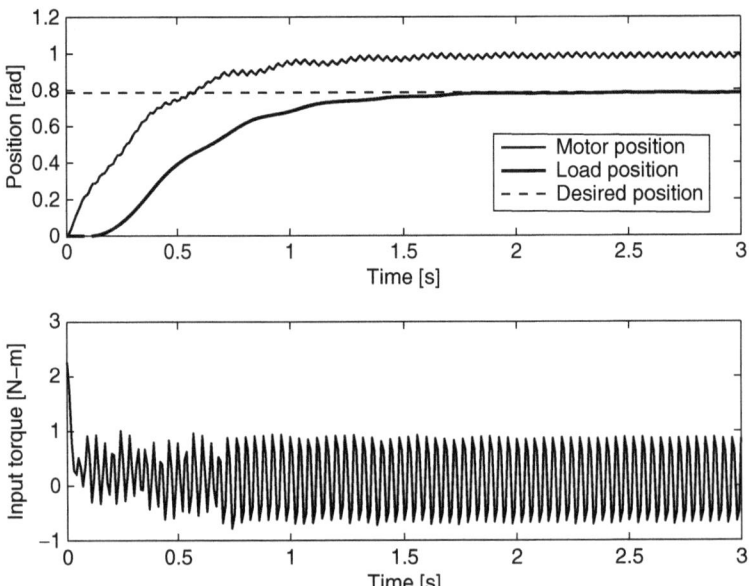

Fig. 9.5 Experimental results for the $\mathcal{H}_\infty$ regulator considering only the motor position information available for feedback: the perturbed case

9.2 Output Regulation of a Coal-Fired Boiler/Turbine Unit with Actuator Deadzone

On a number of boiler/turbine units, the turbine valve position actuation for manipulation of the steam flow rate is of a mechanical-hydraulic type, with the valve position controlled by an electric servo motor, similar to that of Sect. 9.1. The latter is positioned by a device called a motor-driven actuator (MDACT), which is itself a closed-loop system. The MDACT accepts as input the command from the controller of the plant and generates pulses designed to move a motor in small increments to drive the turbine valve to the desired position. Like all pulse-type controllers, the MDACT must have a dead band to prevent the outputs from continuously moving up and down and creating excessive actuator activity. Whenever the actual turbine valve position comes within a small neighborhood of the turbine valve control command, the MDACT stops driving the turbine valve. As a result, there is always a small mismatch between the turbine valve command and its actual position, producing a small steady-state regulation error in the plant outputs.

Consistently addressing such nonidealities, while also providing robust performance, has been a longstanding problem in the control of such systems. A coal-fired boiler/turbine unit's robust performance has been attained, in part, through the linear $\mathcal{H}_\infty$ design reported in [20, 140] as well as in other publications. However, a longstanding problem with the standard linear $\mathcal{H}_\infty$ controllers designed to provide zero steady-state error regulation, such as those in [20, 140], is that they always try to drive this error to zero. This leads to the undesirable oscillations in the control signals and the plant outputs. This problem was addressed in [20] through heuristic modification of the linear $\mathcal{H}_\infty$ topology, referred to as the setpoint conditioning, at the expense of a partial loss of stability robustness, although still outperforming the existing PID-based unit controller. Following [19], this problem is addressed head on in a consistent manner, namely, through the nonsmooth autonomous $\mathcal{H}_\infty$-output controller synthesis of Sect. 6.2, which specializes by taking into account the dead zone in the transmitted actuator signal, retaining the performance robustness of the $\mathcal{H}_\infty$ approach. The efficacy of the proposed synthesis is illustrated through application to the experimentally obtained boiler/turbine model.

9.2.1 Identified Model of a Coal-Fired Boiler/Turbine Unit

A boiler/turbine unit considered in this chapter is the coal-fired boiler/turbine Unit 9, one of the main units at Kingston Fossil Plant located in Kingston, TN. This plant produces approximately 10 billion kilowatt hours of electricity each year and is among the largest in TVA's family of coal-fired plants.

Figure 9.6 shows the closed-loop diagram of the control of the boiler/turbine unit, where the two control inputs are the firing rate and the turbine valve position

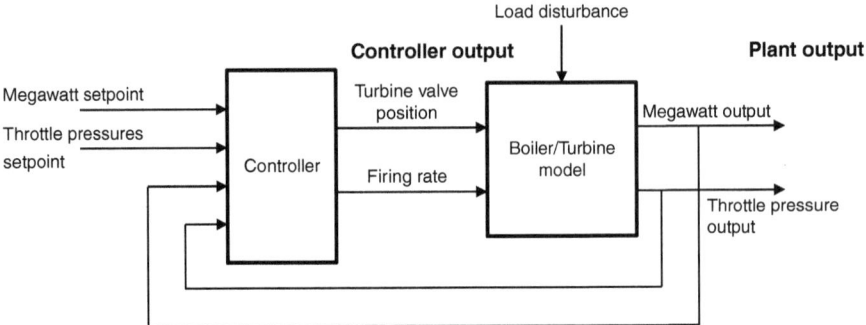

Fig. 9.6 Closed-loop block diagram of a controlled boiler/turbine unit

and the two plant outputs are the megawatt output and the throttle pressure. The control objective is to maintain setpoints under the load and other disturbances and to provide smooth and fast tracking response to the setpoint change, with response characteristics insensitive to varying conditions such as changing coal properties and buildup of soot deposits.

Our specific robust control objectives are

1. to guarantee stable steady-state operation around operating points (185 MW, 1,820 psi) and (150 MW, 1,820 psi),
2. to guarantee fast tracking of MW reference input in the range between 150–185 MW, while keeping the throttle pressure at 1,820 psi with overshoot or undershoot less than 15 psi in magnitude.

The matrix transfer function of a coal-fired boiler/turbine unit, taken from [20], is given by

$$\begin{bmatrix} \theta_1(s) \\ \theta_2(s) \end{bmatrix} = \begin{bmatrix} G_{11}(s) & G_{12}(s) \\ G_{21}(s) & G_{22}(s) \end{bmatrix} \begin{bmatrix} v_1(s) \\ v_2(s) \end{bmatrix}, \qquad (9.16)$$

where $G_{11}(s)$, $G_{12}(s)$, $G_{21}(s)$, and $G_{22}(s)$ are strictly proper stable transfer functions given in (9.18). Here, θ_1 and θ_2 are the megawatt and throttle pressure outputs, respectively; v_1 and v_2 are the turbine valve position and firing rate delivered by actuators whose dynamics is given by

$$J\ddot{q} + f_v \dot{q} + v = \tau_m + w_2, \qquad (9.17)$$

where $q(t) \in \mathbb{R}^2$, $\dot{q}(t) \in \mathbb{R}^2$, and $\ddot{q}(t) \in \mathbb{R}^2$ denote, respectively, the position, velocity, and acceleration of the actuator; $\tau_m \in \mathbb{R}^2$ stands for the input torque, the inertia matrix $J = \text{diag}\{J_1, J_2\}$ is positive definite, the viscous friction coefficients $f_v = \text{diag}\{f_{v1}, f_{v2}\}$ are positive definite matrices of appropriate dimensions, and $w_2(t) \in \mathbb{R}^2$ is introduced to account for destabilizing hard-to-model nonlinear phenomena such as friction and backlash.

9.2 Output Regulation of a Coal-Fired Boiler/Turbine Unit with Actuator Deadzone

$$G_{11}(s) = \frac{-0.0025291s(s-2.12)(s+0.1783)(s+0.06005)(s+0.01543)}{(s+0.195)(s+0.0752)(s^2+0.02651s+0.0004476)(s^2+0.08533s+0.002877)},$$

$$G_{12}(s) = \frac{-0.00092731(s-2.136)(s^2+0.07913s+0.002069)(s^2-0.09745s+0.01647)}{(s+0.195)(s+0.0752)(s^2+0.02651s+0.0004476)(s^2+0.08533s+0.002877)},$$

$$G_{21}(s) = \frac{-0.0014435(s+0.1478)(s^2+0.05292s+0.001739)(s^2-0.7298s+0.7524)}{(s+0.195)(s+0.0752)(s^2+0.02651s+0.0004476)(s^2+0.08533s+0.002877)},$$

$$G_{22}(s) = \frac{-0.00066211(s-5.62)(s^2+0.08217s+0.002909)(s^2-0.2491s+0.05502)}{(s+0.195)(s+0.0752)(s^2+0.02651s+0.0004476)(s^2+0.08533s+0.002877)}.$$

(9.18)

Here, we consider that the valve position v_1 and the firing rate v_2 have dead zone effects, the former due to MDACT described in Sect. 9.2.1 and the latter because of the coal feeder's inability to exactly match firing rate demand, modeled by (see (9.2) for the one-dimensional dead zone model)

$$v_i = \begin{cases} 0 & \text{if } |\Delta y_i| \leq j_i \\ K_i \Delta y_i - K_i j_i \operatorname{sign}(\Delta y_i) & \text{otherwise} \end{cases}, \quad i = 1, 2, \qquad (9.19)$$

where the positive constants K_i and j_i stand for stiffness and amplitude, respectively, and

$$\Delta y_i = q_i - N_i \theta_i, \quad i = 1, 2, \qquad (9.20)$$

N_1 stands for the ratio rate in rad/MW, and N_2 for the ratio rate in rad/psi.

9.2.2 Problem Statement

To formally state the problem, let's consider the state-space representation of (9.16)–(9.18) augmented with a disturbance vector $w_1(t) \in \mathbb{R}^2$:

$$\dot{x}_p = A_p x_p + B_p w_1 + B_p v, \qquad (9.21)$$

$$\theta = C_p x_p, \qquad (9.22)$$

where $x_p = [x_1, \ldots, x_{12}]^T$ is the state vector and matrices

$$A_p = \begin{bmatrix} A_{1p} & 0_{6\times 6} \\ 0_{6\times 6} & A_{2p} \end{bmatrix}, \quad B_p = \begin{bmatrix} B_{1p} \\ B_{2p} \end{bmatrix}, \quad C_p = \begin{bmatrix} C_{1p} & C_{2p} \end{bmatrix} \qquad (9.23)$$

are specified as follows:

$$A_{1p} = \begin{bmatrix} -0.1950 & 0 & 0 & 0 & 0 & 0 \\ 0 & -0.1950 & 0 & 0 & 0 & 0 \\ 0 & 0 & -0.0133 & 0.0165 & 0 & 0 \\ 0 & 0 & -0.0165 & -0.0133 & 0 & 0 \\ 0 & 0 & 0 & 0 & -0.0133 & 0.0165 \\ 0 & 0 & 0 & 0 & -0.0165 & -0.0133 \end{bmatrix},$$

$$A_{2p} = \begin{bmatrix} -0.0427 & 0.0325 & 0 & 0 & 0 & 0 \\ -0.0325 & -0.0427 & 0 & 0 & 0 & 0 \\ 0 & 0 & -0.0427 & 0.0325 & 0 & 0 \\ 0 & 0 & -0.0325 & -0.0427 & 0 & 0 \\ 0 & 0 & 0 & 0 & -0.0752 & 0 \\ 0 & 0 & 0 & 0 & 0 & -0.0752 \end{bmatrix},$$

$$B_{1p}^T = \begin{bmatrix} 10.2225 & -7.0847 & -548.1562 & -44.9752 & 38.0543 & 477.0014 \\ -0.5487 & 5.5395 & -94.5313 & -94.5011 & 104.6406 & 86.2761 \end{bmatrix},$$

$$B_{2p}^T = \begin{bmatrix} 104.5026 & 506.7193 & -230.8045 & -94.1148 & 237.4779 & 48.8719 \\ -65.5447 & 132.4462 & -114.9810 & 117.7746 & 101.2994 & -115.4110 \end{bmatrix},$$

$$C_{1p} = \begin{bmatrix} -0.0059 & -0.0079 & -0.0006 & 0.0000 & -0.0002 & -0.0007 \\ -0.0208 & -0.0273 & -0.0059 & 0.0017 & -0.0038 & -0.0061 \end{bmatrix},$$

$$C_{2p} = \begin{bmatrix} 0.0003 & -0.0002 & -0.0004 & -0.0003 & 0.0000 & -0.0009 \\ -0.0001 & 0.0007 & 0.0008 & -0.0002 & 0.0000 & -0.0058 \end{bmatrix}.$$

(9.24)

The above space-state representation was derived using the MATLAB® commands ssdata for obtaining the state matrices, minreal for the minimal realization, and canon for diagonalization of the matrix A_p. For the dynamics of the actuators, one has

$$\ddot{x}_a = J^{-1}[\tau_m - f_v \dot{x}_a - v + w_2], \qquad (9.25)$$

where $x_a = q$. For convenience, $v \in \mathbb{R}^2$ is represented in terms of the state $x = [x_p, x_a, \dot{x}_a]^T$, thereby yielding

$$v_i(x) = \begin{cases} 0 & \text{if } |\Delta x_i| \leq j_i \\ K_i \Delta x_i - K_i j_i \operatorname{sign}(\Delta x_i) & \text{otherwise} \end{cases}, \quad i = 1, 2, \qquad (9.26)$$

9.2 Output Regulation of a Coal-Fired Boiler/Turbine Unit with Actuator Deadzone

where

$$\Delta x_i = x_{a_i} - N_i \theta_i. \tag{9.27}$$

Equations (9.26)–(9.27) result from the substitution of (9.22) into (9.19) and (9.20).

Taking into account the dynamics of the actuators, we see that the general control objective of the output regulation of the coal-fired power plant (9.21)–(9.27) is to design an $\mathcal{H}_\infty$ controller so as to obtain the closed-loop system in which all the trajectories are bounded and the outputs $[\theta_1, \theta_2]$ asymptotically decay to a desired setpoint $[\theta_{1d}, \theta_{2d}]$ as $t \to \infty$ while also attenuating the influence of the external disturbances $w_1(t), w_2(t)$. It is intended to design a regulator of the form

$$\tau_m = -K_a x_a + u, \tag{9.28}$$

with a diagonal positive definite matrix K_a that imposes the desired stability properties on the disturbance-free system motion while also locally attenuating the effect of the disturbances. Thus, the regulator to be constructed consists of a proportional part and a disturbance attenuator $u \in \mathbf{R}^2$, internally stabilizing the closed-loop system around the desired position.

In the sequel, the investigation is confined to the megawatt output and pressure regulation problem, where

1. The output to be controlled is given by

$$z = \begin{bmatrix} u \\ \rho C_p x_p \end{bmatrix} \tag{9.29}$$

with a positive weight coefficient ρ.

2. The measurements at the side of the actuators $y = x_a \in \mathbb{R}^2$ are the only available, and these measurements are corrupted by the error vector $w_y(t) \in \mathbb{R}^2$; that is,

$$y = x_a + w_y(t). \tag{9.30}$$

The $\mathcal{H}_\infty$-control problem in question is thus stated as follows. Given the system representation (9.21)–(9.30) and a real number $\gamma > 0$, one is required to find (if any) a causal dynamic feedback controller

$$u = \mathcal{K}(\xi), \quad \dot{\xi} = \mathcal{F}(y, \xi), \tag{9.31}$$

with internal state $\xi \in \mathbb{R}^{16}$ such that the undisturbed closed-loop system is uniformly asymptotically stable around the origin and its L_2 gain is locally less than γ; that is, inequality

$$\int_0^T \|z(t)\|^2 dt \le \gamma^2 \int_0^T \|w(t)\|^2 dt \tag{9.32}$$

is satisfied for all $T > 0$ and all piecewise-continuous functions $w(t) = [w_1(t), w_2(t), w_y(t)]^T$ for which the corresponding state trajectory of the closed-loop system, initialized at the origin, remains in some neighborhood of this point.

9.2.3 Nonsmooth $\mathcal{H}_\infty$ Synthesis

To begin, we represent the plant equations (9.21)–(9.27) in the generic form (6.1)–(6.3) by letting $x = [x_p, x_a, \dot{x}_a]^T$ and setting

$$f_1(x) = \begin{bmatrix} A_p x_p \\ \dot{x}_a \\ -K_a x_a - J^{-1} f_v \dot{x}_a \end{bmatrix}, \tag{9.33}$$

$$f_2(x) = \begin{bmatrix} B_p v(x) \\ 0 \\ -J^{-1} v(x) \end{bmatrix} \tag{9.34}$$

$$g_1(x) = \begin{bmatrix} B_p & 0_{12 \times 2} & 0_{12 \times 2} \\ 0_{2 \times 2} & J^{-1} & 0_{2 \times 2} \\ 0_{2 \times 2} & 0_{2 \times 2} & 0_{2 \times 2} \end{bmatrix}, \quad g_2(x) = \begin{bmatrix} 0_{14 \times 2} \\ J^{-1} \end{bmatrix},$$

$$h_1(x) = \rho \begin{bmatrix} 0_{2 \times 1} \\ C_p x_p \end{bmatrix}, \quad h_2(x) = x_a + q_d,$$

$$k_{12}(x) = \begin{bmatrix} I_2 \\ 0_{2 \times 2} \end{bmatrix}, \quad k_{21} = \begin{bmatrix} 0_{2 \times 4} & I_2 \end{bmatrix}. \tag{9.35}$$

Hereinafter, I_n and $0_{n \times m}$ stand for the $n \times n$ identity matrix and the $n \times m$ matrix of zeros, respectively.

Since the plant representation contains the nonsmooth term $f_2(x)$, the present $\mathcal{H}_\infty$-control problem is resolved within the framework of the nonsmooth autonomous $\mathcal{H}_\infty$ synthesis of Sect. 6.2 by applying Theorem 26. To make sure that Theorem 26 is applicable to the above system, it suffices to straightforwardly inspect that the nonsmooth term $f_2(x)$ is locally Lipschitz continuous in x around the origin $x = 0$ and to establish that the vector $\zeta = 0$ is a proximal supergradient of the components $f_{2i}(x)$ ($i = 1, \ldots, n$) of the function $f_2(x)$ at $x = 0$. The following result is thus in order.

Theorem 37. *Let Conditions $C1''$ and $C2''$ of Sect. 6.2 hold for the matrix functions A, B_1, B_2, C_1, C_2, governed by (6.35), (9.33)–(9.35), and let $(P_\varepsilon, Z_\varepsilon)$ be the corresponding positive definite solution of (6.64), (6.65) under some $\varepsilon > 0$. Then the output feedback*

$$\dot{\xi} = f_1(\xi) + f_2(\xi) + \left[\frac{1}{\gamma^2}g_1(\xi)g_1^T(\xi) - g_2(\xi)g_2^T(\xi)\right]P_\varepsilon\xi$$
$$+ Z_\varepsilon C_2^T[y - C_2\xi], \tag{9.36}$$
$$u = -g_2^T(\xi)P_\varepsilon\xi, \tag{9.37}$$

subject to (9.33)–(9.35), is a local solution of the $\mathcal{H}_\infty$-control problem for the coal-fired boiler/turbine unit (9.16)–(9.18).

Proof. As in the proof of Theorem 36, all the assumptions of Sect. 6.2 are trivially verified for system (6.1)–(6.3), specified with (9.33)–(9.35). Thus, Theorem 26 proves to be applicable to the system in question. By applying Theorem 26 to (6.1)–(6.3) subject to (9.33)–(9.35), we've then straightforwardly established the validity of the present theorem.

9.2.4 Simulation Results

The performance of the closed-loop system was verified by simulation using MATLAB®. The parameters selected for the controller were $\gamma = 30$, $\rho = 1$, and $\varepsilon = 0.001$. The dead zone amplitudes for (9.19) were $j_1 = 0.1$ and $j_2 = 0.7$. For the actuators, we chose $J = \mathrm{diag}\{0.01, 0.01\}$, $f_v = \mathrm{diag}\{0.1, 0.1\}$, $N_i = 1$, and $K_i = 10$ ($i = 1, 2$). For the proportional part, we set $K_a = \mathrm{diag}\{10, 10\}$.

We set $\theta_{1d} = 185\,\mathrm{MW}$ and $\theta_{2d} = 1{,}820\,\mathrm{psi}$ as the desired megawatt and throttle pressure, respectively. The initial conditions selected for the simulations were $\theta_1(0) = 180\,\mathrm{MW}$ and $\theta_2(0) = 1{,}805\,\mathrm{PSI}$.

The good performance of the closed-loop system is concluded from Fig. 9.7, which depicts the megawatt output θ_1 and throttle pressure θ_2. The simulation indicates that the proposed $\mathcal{H}_\infty$ synthesis removes oscillations around the operating point that characterize the existing linear $\mathcal{H}_\infty$-design results.

In order to illustrate attractive robustness features of the proposed synthesis, we realistically assumed the real size of the dead zone would be unavailable to the designer. Just in case, the parameters $j_1 = 0.1$ and $j_2 = 0.7$, still used in the dead zone model of the plant (9.21)–(9.22), mismatched to the parameters $j_1 = 0.4$ and $j_2 = 1.4$ used in the dead zone model (9.34) of compensator (9.36). Figure 9.8 shows the output responses and control inputs of the closed-loop system, driven by the mismatched regulator (9.37) thus constructed. From this figure, one may conclude the strong robustness of the proposed synthesis against the dead zone variation.

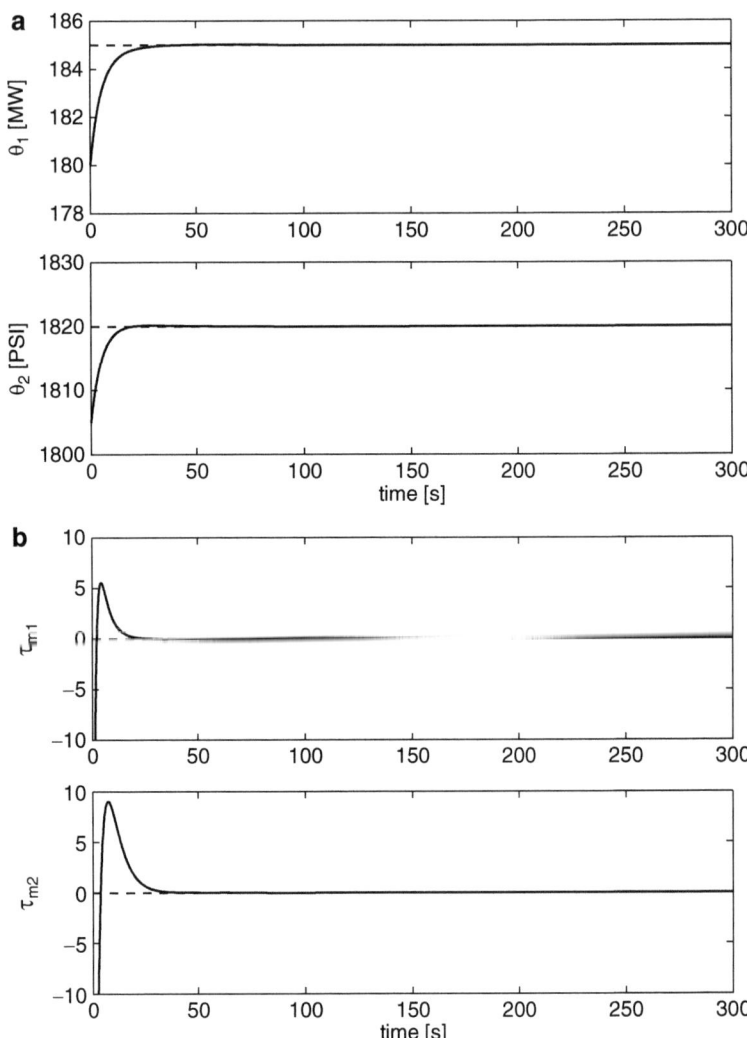

Fig. 9.7 Output responses (**a**) and input control (**b**) of the closed-loop system

9.2 Output Regulation of a Coal-Fired Boiler/Turbine Unit with Actuator Deadzone

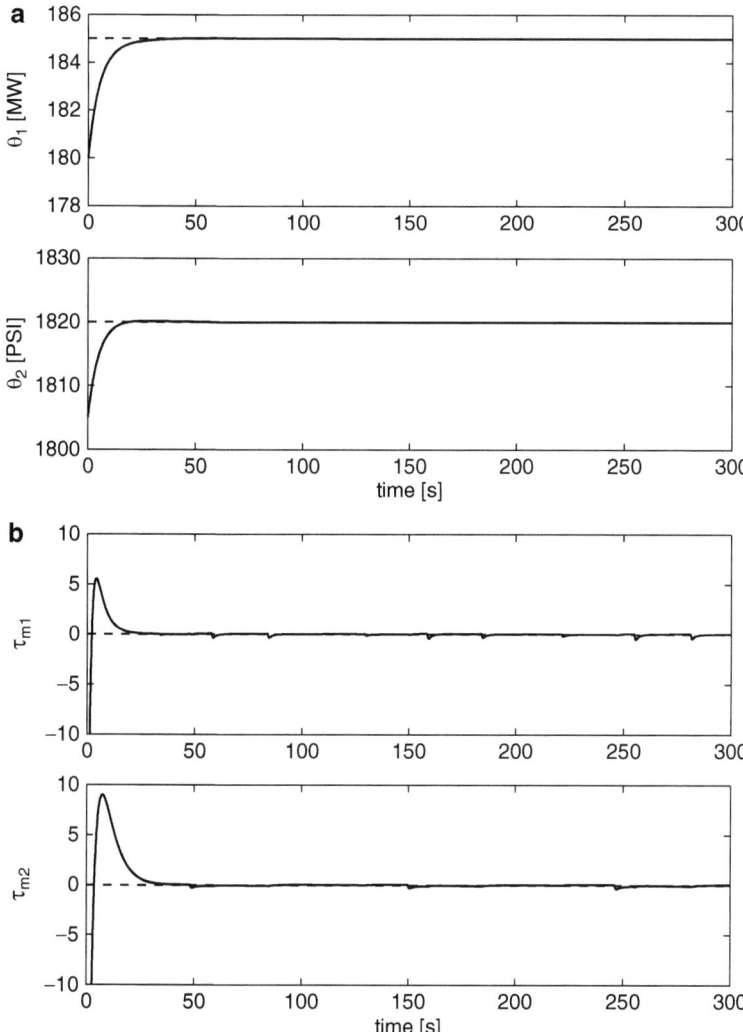

Fig. 9.8 Output responses (**a**) and control input (**b**) of the closed-loop system under the assumption that the dead zone is not exactly known

Chapter 10
$\mathcal{H}_\infty$ Generation of Periodic Motion of Mechanical Systems of One Degree of Underactuation

As seen in Chap. 8, the motion-planning problem for revolute joint robotic systems, whose number of actuators corresponds to its number of degrees of freedom, consists of specifying a set of trajectories in the configuration space of the robot. The configuration of a manipulator is a complete specification of the robot's location. The set of all possible configurations is typically referred to as the configuration space. The key point is that given any prespecified trajectory to follow, a global tracking controller can be designed and decoupled in n components, separately driving each of the n joints of the robot, as in the case of the feedback linearization [115].

In contrast to fully actuated systems, in underactuated systems, the presence of unactuated links, whose dynamics depend on those of actuated links, does not allow one to independently control each link of such a system. A powerful analytical tool of generating a periodic motion in underactuated mechanical systems relies on the virtual constraint approach [112] to subsequently be presented in terms of the joint generalized coordinates. Although the configuration is generally given in Cartesian coordinates and orientation, however, using inverse kinematics, they can be transformed into joint coordinates. Coupled with the virtual constraint approach, the $\mathcal{H}_\infty$ generation of periodic motion is further developed and applied to a 3-DOF helicopter prototype of one degree of underactuation.

To facilitate the exposition, the present study focuses on smooth underactuated systems, whereas the nonsmooth $\mathcal{H}_\infty$ synthesis of Chap. 6.1 is still in play for its potential capability of attenuating nonsmooth effects. The extension of the virtual constraint approach to periodic motion generation in the presence of nonsmooth phenomena (such as dry friction forces and backlash as well as collisions) is indeed of practical interest. While being an open problem, such an extension calls for a separate investigation that remains beyond the present development.

10.1 Virtual Constraint Approach

The *virtual holonomic constraint* approach is a power analytic tool of planning periodic motion in underactuated systems (see, e.g., [26,46,47,111]). The main idea of this approach is to generate a simplest lower-dimension system, called the *virtual limit system*, by specifying desired geometric relations among the generalized coordinates of the mechanism and impose them by feedback control action. These imposed constraints are called virtual since they are forced by an appropriate feedback strategy and are not physically presented in the mechanism.

The approach is subsequently detailed via a controlled n-DOF Euler–Lagrange system of underactuation degree 1:

$$\frac{d}{dt}\left(\frac{\partial \mathcal{L}(q,\dot{q})}{\partial \dot{q}}\right) - \frac{\partial \mathcal{L}(q,\dot{q})}{\partial q} = B(q)u. \tag{10.1}$$

Hereinafter, $q(t) = (q_1(t),\ldots,q_n(t))^T$ and $\dot{q}(t) = (\dot{q}_1(t),\ldots,\dot{q}_n(t))^T$ are vectors of generalized coordinates and velocities, $u(t) = (u_1(t),\ldots,u_{n-1}(t))^T$ is a vector of control inputs, and $\mathcal{L}(q,\dot{q})$ is the Lagrangian given by

$$\mathcal{L}(q,\dot{q}) = \frac{1}{2}\dot{q}^T M(q)\dot{q} - V(q), \tag{10.2}$$

where $M(q) \in \mathbb{R}^{n\times n}$ is the inertia matrix, which is positive definite everywhere, $V(q)$ denotes the potential energy, and $B(q) \in \mathbb{R}^{n\times(n-1)}$ is a matrix of full rank; the matrix functions $M(q), V(q), B(q)$ are normally of class C^1. Without loss of generality, it is assumed that the first $n-1$ generalized coordinates $q_1,\ldots,q_{n-1}$ are straightforwardly actuated by the control input $u(\cdot)$, whereas q_n remains unactuated. It should be noted that the extension of the approach to Lagrangian systems of a higher underactuation degree is not straightforward, and it has not been attained in the literature at the same level of generality as has been done for one degree of underactuation.

A trajectory $q(t)$ in the $2n$-dimensional state space of the Euler–Lagrange system (10.1) is represented by a solution of (10.1) initialized at $q(0) = q_0, \dot{q}(0) = \dot{q}_0$. Motion planning aims to achieve a certain task by carrying out a feasible trajectory $q_\star(t)$ of the Euler–Lagrange system (10.1) driven by a feedback signal $u_\star(q,\dot{q})$ of class C^1 such that

$$|\dot{q}_\star(t)|^2 + |\ddot{q}_\star(t)|^2 > 0$$

for all $t \in [0,T]$. Hereinafter, $T > 0$ stands for the duration of the motion, and it is not necessarily finite. The orbit of the corresponding trajectory is

$$\mathcal{O}_\star = \{(q,\dot{q}) \in \mathbb{R}^{2n} : q = q_\star(\tau), \dot{q} = \dot{q}_\star(\tau), \tau \in [0,T]\} \tag{10.3}$$

10.1 Virtual Constraint Approach

and its tubular δ-neighborhood is given by

$$\mathcal{O}_\delta(q_\star) = \left\{ (q,\dot{q}) : \min_{t \in [0,T]} \left\| \begin{bmatrix} q - q_\star(\tau) \\ \dot{q} - \dot{q}_\star(\tau) \end{bmatrix} \right\| \le \delta \right\}, \tag{10.4}$$

with some $\delta > 0$.

10.1.1 Reparametrization of a Motion

According to [7], system (10.1) is representable in the form

$$M(q)\ddot{q} + C(q,\dot{q})\dot{q} + G(q) = B(q)u, \tag{10.5}$$

where $C(q,\dot{q}) = \frac{\partial(M(q)\dot{q})}{\partial q} \in \mathbb{R}^{n \times n}$ is the matrix of Coriolis and centrifugal forces, and $G(q) = [(\partial V(q)/\partial q_1), \ldots, (\partial V(q)/\partial q_n)]^T$ is a vector of gravitational forces.

It suffices to describe a solution of system (10.5) in terms of the evolution of the generalized coordinates

$$q_1 = q_1(t), \ldots, q_n = q_n(t), \qquad t \in [0,T], \tag{10.6}$$

as functions of time because the complement of the entire state vector consists of the generalized velocities, being the time derivatives of these functions.

Alternatively, a time-independent description, such as the orbital representation (10.3), is available in a parametric form via a set of geometric relations

$$q_1 = f_1(\phi), \ldots, q_n = f_n(\phi), \tag{10.7}$$

valid along the orbit, where $f_i(\phi)$, $i = 1, \ldots, n$, are functions of a parameter ϕ. Thus, a geometric description of the path is given in terms of the parameter $\phi(t)$, evolving along with $t \in [0,T]$. Identities (10.7) are known as *virtual holonomic constraints* since they express algebraic relations among the generalized coordinates, whereas certain relations among the generalized velocities follow from the coordinate constraints.

Once the parameter ϕ is chosen [in many cases, one of the coordinates, say $q_n = \phi$, is a good parameter choice with $f_n(\phi) = \phi$], the dynamics (10.5) in the new coordinates (10.7) are obtained by substituting the time derivatives $\dot{q}_i = f_i'\dot{\phi}$, $\ddot{q}_i = f_i''\dot{\phi}^2 + f_i'\ddot{\phi}$, $i = 1, \ldots, n$, of (10.7) into (10.5), where $f_i' = \partial f_i/\partial \phi$ and $f_i'' = \partial^2 f_i/\partial \phi^2$. When one takes into account that $C(q,\dot{q}) = \partial(M(q)\dot{q})/\partial q$ is linear in $\dot{q}$, the resulting equation is then governed by

$$M(F(\phi))[F''\dot{\phi}^2 + F'\ddot{\phi}] + C(F(\phi), F')F'\dot{\phi}^2 + G(F(\phi)) = B(F(\phi))u, \tag{10.8}$$

where

$$F(\phi) = [f_1(\phi), \ldots, f_n(\phi)]^T,$$
$$F'(\phi) = [f'_1(\phi), \ldots, f'_n(\phi)]^T,$$
$$F''(\phi) = [f''_1(\phi), \ldots, \phi''_n(\phi)]^T. \tag{10.9}$$

Since the present development is confined to controlled dynamics (10.5) with one degree of underactuation, there exists a nontrivial matrix function $B^\perp(q) \in \mathbb{R}^{1 \times n}$ such that $B^\perp(q)B(q) = 0$. Thus, multiplying (10.8) by $B^\perp(q)$ from the left, one arrives at the reduced second-order dynamics along the holonomic constraints (10.7):

$$\alpha(\phi)\ddot{\phi} + \beta(\phi)\dot{\phi}^2 + \gamma(\phi) = 0, \tag{10.10}$$

where

$$\alpha(\phi) = B^\perp(F(\phi))M(F(\phi))F'(\phi),$$
$$\beta(\phi) = B^\perp(F(\phi))[C(F(\phi), F')F' + M(F(\phi))F''(\phi)],$$
$$\gamma(\phi) = B^\perp(F(\phi))G(F(\phi)). \tag{10.11}$$

To summarize, the following result is in order.

Lemma 11 ([110]). *Consider the n-DOF controlled system (10.5) with $n-1$ independent control inputs, that is, of underactuation degree 1. Let $q = q(t)$ be a motion (10.6) of (10.5) in response to a control signal $u = u(t)$, both defined on the time interval $t \in [0, T]$. Let $\phi = \phi$ be a scalar variable used in (10.7) to parameterize the motion $q(t)$. Then $\phi(t)$ is not arbitrary, but it is a solution of the second-order nonlinear differential equation (10.10) with some smooth scalar functions $\alpha(\phi)$, $\beta(\phi)$, and $\gamma(\phi)$.*

Thus, given a virtual constraint (10.7), we see that the dynamics (10.6) of (10.5) possess a periodic motion whenever the corresponding differential equation (10.10) exhibits a periodic solution. If specified with $\alpha(\phi) = 1$, $\beta(\phi) = 0$, and $\gamma(\phi) = \omega^2 \phi$, Eq. (10.10) presents a simple harmonic oscillator with periodic solutions $\phi(t) = k\cos(\omega t + f)$ of frequency $\omega > 0$, arbitrary amplitude $k > 0$, and arbitrary phase $f > 0$.

General sufficient conditions for the second-order differential equation (10.10) to have a periodic solution can be derived from an explicit representation of an integral $I(\phi, \dot{\phi}, \phi_0, \dot{\phi}_0)$ of the motion being a function of the arguments $\phi, \dot{\phi}, \phi_0, \dot{\phi}_0$ with a permanent value along a solution $\phi = \phi(t), \dot{\phi} = \dot{\phi}(t)$ initialized with $\phi(0) = \phi_0, \dot{\phi}(0) = \dot{\phi}_0$. Such an integral is obtained based on the transformation of the second-order differential equation (10.10) with respect to the time variable t into the first-order differential equation

$$\frac{1}{2}\alpha(\phi)\frac{dY}{d\phi} + \beta(\phi)Y + \gamma(\phi) = 0 \tag{10.12}$$

10.1 Virtual Constraint Approach

with respect to the variable ϕ, where ϕ serves as internal timing in (10.5) and

$$Y(\phi(t)) = \dot{\phi}^2(t). \tag{10.13}$$

To reproduce the above equation from (10.10), we must note that

$$\frac{d^2}{dt^2}\phi(t) = \frac{d}{dt}(\dot{\phi}(t)) = \frac{d}{d\phi}(\dot{\phi}(t))\frac{d}{dt}\phi(t) = \frac{d}{d\phi}(\dot{\phi}(t))\dot{\phi}(t) = \frac{d}{d\phi}\left(\frac{1}{2}\dot{\phi}^2(t)\right).$$

Provided that

$$\alpha(\phi(t)) \neq 0 \text{ for all } t \in [0, T], \tag{10.14}$$

the general solution of (10.12) has the form

$$Y(\phi) = \Psi(\phi, \phi_0)Y(\phi_0) - \Psi(\phi, \phi_0)\int_{\phi_0}^{\phi} \Psi(\phi_0, s)\frac{2\gamma(s)}{\alpha(s)}ds, \tag{10.15}$$

where

$$\Psi(\phi_1, \phi_0) = \exp\left\{-\int_{\phi_0}^{\phi_1} \frac{2\beta(\tau)}{\alpha(\tau)}d\tau\right\}. \tag{10.16}$$

Substituting (10.13) into (10.15) yields

$$\dot{\phi}^2(t) = \Psi(\phi, \phi_0)\dot{\phi}^2(0) - \Psi(\phi, \phi_0)\int_{\phi_0}^{\phi} \Psi(\phi_0, s)\frac{2\gamma(s)}{\alpha(s)}ds, \tag{10.17}$$

thereby ensuring the following result.

Lemma 12 ([113]). *Provided that Condition (10.14) holds on the solution domain $[0, T]$, the integral function*

$$I(\phi, \dot{\phi}, \phi_0, \dot{\phi}_0) = \dot{\phi}^2 - \Psi(\phi, \phi_0)\left[\dot{\phi}_0^2 - \int_{\phi_0}^{\phi} \Psi(\phi_0, s)\frac{2\gamma(s)}{\alpha(s)}ds\right], \tag{10.18}$$

where $\Psi(\phi_1, \phi_0)$ is given by (10.16), preserves the zero value along the solutions $(\phi(t), \dot{\phi}(t))$ of (10.10), initialized with $\phi(0) = \phi_0, \dot{\phi}(0) = \dot{\phi}_0$.

An analytical tool of detecting cycles in system (10.10) has been established, based on the knowledge of the integral function (10.18).

Theorem 38 (Sufficient conditions for the existence of periodic solutions [113]). *Let ϕ_0 be an equilibrium point of system (10.10), namely, $\gamma(\phi_0) = 0$. Suppose that*

1. *The scalar functions $\alpha(\phi)$, $\beta(\phi)$, and $\gamma(\phi)$ are locally continuous around ϕ_0.*
2. *The fractional function $\gamma(\phi)/\alpha(\phi)$ is continuously differentiable locally around at $\phi = \phi_0$.*
3. *The nonlinear system (10.10) is locally well posed around $(\phi = \phi_0, \dot{\phi} = 0)$.*

Then the nonlinear system (10.10) has a center at the equilibrium ϕ_0 whenever the linearized system

$$\ddot{z} + \left(\frac{d}{d\phi}\frac{\gamma(\phi)}{\alpha(\phi)}\bigg|_{\phi=\phi_0}\right)z = 0 \tag{10.19}$$

has a center at $z = 0$.

10.1.2 Coordinate Transformation

Let $q_\star(t)$ be a feasible trajectory of system (10.5) driven by a feedback signal $u_\star(q, \dot{q})$ of class C^1. Given the virtual constraints (10.7), one can view the quantities

$$\phi, \quad y_1 = q_1 - f_1(\phi), \ldots, y_n = q_n - f_n(\phi) \tag{10.20}$$

as redundant generalized coordinates for the underactuated system (10.5) so that one of them, say y_n, can be expressed as a function of the others. Then new independent coordinates are

$$y = (y_1, \ldots, y_{n-1})^T \in \mathbb{R}^{n-1} \text{ and } \phi \in \mathbb{R}, \tag{10.21}$$

whereas the last equality of (10.20) can be locally rewritten as

$$q_n = f_n(\phi) + h(y, \phi), \tag{10.22}$$

where $h(y, \phi)$ is a smooth scalar function of its arguments. The coordinate transformation (10.20), (10.22) comes with the Jacobian matrix

$$L(\phi, y) = \begin{bmatrix} 1_{n-1} & 0_{(n-1)\times 1} \\ \frac{\partial h}{\partial y} & \frac{\partial h}{\partial \phi} \end{bmatrix} + \begin{bmatrix} 0_{n\times(n-1)} & \begin{array}{c} \frac{\partial \phi_1(\phi)}{\partial \phi} \\ \vdots \\ \frac{\partial \phi_n(\phi)}{\partial \phi} \end{array} \end{bmatrix}, \tag{10.23}$$

where $1_{n-1} \in \mathbb{R}^{(n-1)\times(n-1)}$ is the identity matrix, and $0_{(n-1)\times 1} \in \mathbb{R}^{(n-1)\times 1}$ and $0_{n\times(n-1)} \in \mathbb{R}^{n\times(n-1)}$ are the matrices with zero entries. Provided that the Jacobian matrix (10.23) is nonsingular (it is the case, e.g., when $q_n = \phi$), the one-to-one relation is locally established for the remaining state variables, namely, between the first-order derivatives of the new coordinates $(y, \phi)^T$ and those of the original coordinates q:

$$\dot{q} = L(y, \phi) \begin{bmatrix} \dot{y} \\ \dot{\phi} \end{bmatrix}. \tag{10.24}$$

10.1 Virtual Constraint Approach

It follows that

$$\ddot{q} = L(y,\phi)\begin{bmatrix}\ddot{y}\\\ddot{\phi}\end{bmatrix} + \dot{L}(y,\phi)\begin{bmatrix}\dot{y}\\\dot{\phi}\end{bmatrix}, \qquad (10.25)$$

where $\dot{L}(y,\phi) = dL(y,\phi)/dt$. By employing (10.5), we can represent the above relation in the form

$$\ddot{q} = M^{-1}(q)(B(q)u - C(q,\dot{q})\dot{q} - G(q)) = L(y,\phi)\begin{bmatrix}\ddot{y}\\\ddot{\phi}\end{bmatrix} + \dot{L}(y,\phi)\begin{bmatrix}\dot{y}\\\dot{\phi}\end{bmatrix}, \qquad (10.26)$$

resulting in

$$\begin{bmatrix}\ddot{y}\\\ddot{\phi}\end{bmatrix} = L^{-1}(\phi,y)\left[M^{-1}(q)(B(q)u - C(q,\dot{q})\dot{q} - G(q)) - \dot{L}(y,\phi)\begin{bmatrix}\dot{y}\\\dot{\phi}\end{bmatrix}\right]. \qquad (10.27)$$

Substituting $q_i = y_i - f_i(\phi)$, $1 \leq i \leq (n-1)$, $q_n = f_n(\phi) + h(y,\phi)$, and $\dot{q} = L(y,\phi)[\dot{y} \ \dot{\phi}]^T$ into (10.27) yields the state equations governing the dynamics of the variable y. Thus,

$$\ddot{y} = R(y,\dot{y},\phi,\dot{\phi}) + N(y,\phi)u, \qquad (10.28)$$

where

$$R(y,\dot{y},\phi,\dot{\phi}) = \begin{bmatrix}1_{(n-1)} & 0_{(n-1)\times 1}\end{bmatrix} L^{-1}(y,\phi)$$

$$\times \left[M^{-1}(q)(-C(q,\dot{q})\dot{q} - G(q)) - \dot{L}(y,\phi)\begin{bmatrix}\dot{y}\\\dot{\phi}\end{bmatrix}\right],$$

$$N(y,\phi) = \begin{bmatrix}1_{(n-1)} & 0_{(n-1)\times 1}\end{bmatrix} L^{-1}(y,\phi)M^{-1}(q)B(q). \qquad (10.29)$$

10.1.3 Coordinate-Feedback Transformation

Since the system (10.5) has one degree of underactuation and, for certainty, the unactuated coordinate has been denoted by q_n, the matrix $N(y,\phi)$ in (10.29) proves to be invertible by inspection. Hence, the coordinate-feedback transform

$$u = N^{-1}(y,\phi)(u_\star + v) \qquad (10.30)$$

is well posed from u to v, and it brings system (10.28) to the form

$$\ddot{y} = R(y,\dot{y},\phi,\dot{\phi}) + u_\star + v, \qquad (10.31)$$

where each component y_i, $i = 1, \ldots, n-1$, of the generalized coordinate y is manipulated only by the corresponding component v_i of the new control input v, and $u_\star$ is the nominal feedback that drives the system along the feasible trajectory $q_\star(t)$. Another attractive feature of the above representation, which is relevant to the transversal linearization, is that the $(n-1)$-vector function

$$\Phi(y, \dot{y}, \phi, \dot{\phi}) = R(y, \dot{y}, \phi, \dot{\theta}) + u_\star(y, \dot{y}, \phi, \dot{\phi}) \tag{10.32}$$

is nullified for $y = \dot{y} = 0$; that is, $\Phi(0, 0, \phi, \dot{\theta}) = 0$.

To complete the system representation, we need to augment (10.31) with a scalar second-order differential equation for the variable ϕ. Following the same reasoning used for the derivation of Eq. (10.10) along the virtual constraint (10.7) and taking into account that beyond this constraint, relations (10.20) and (10.22) are in force, we find that such a complement equation

$$\alpha(\phi)\ddot{\theta} + \beta(\phi)\dot{\phi}^2 + \gamma(\phi) = g(y, \dot{y}, \theta, \dot{\phi}, v) \tag{10.33}$$

is extracted from (10.27) with the same functions $\alpha(\phi), \beta(\phi), \gamma(\phi)$ as in (10.10) and with a scalar smooth function $g(y, \dot{y}, \phi, \dot{\phi}, \ddot{\phi}, v)$, which is nullified for $y = \dot{y} = 0$. It follows that the order degeneration of Eq. (10.33) is locally impossible under condition (10.14) within a vicinity of the virtual constraint (10.7), where $y = \dot{y} = 0$.

Following [112], (10.33) admits the representation

$$\alpha(\phi)\ddot{\phi} + \beta(\phi)\dot{\phi}^2 + \gamma(\phi) = g_I(y, \dot{y}, \phi, \dot{\phi}, \ddot{\phi})I + g_y(y, \dot{y}, \phi, \dot{\phi}, \ddot{\phi})y$$
$$+ g_{\dot{y}}(y, \dot{y}, \phi, \dot{\phi}, \ddot{\phi})\dot{y} + g_v(y, \dot{y}, \phi, \dot{\phi})v, \tag{10.34}$$

where, along with $y, \dot{y}$, the coordinate I, determined by the integral function (10.18), is transversal to the orbit $q_\star$, and $g_I(\cdot)$, $g_y(\cdot)$, $g_{\dot{y}}(\cdot)$, and $g_v(\cdot)$ are smooth matrix functions of appropriate dimensions.

Theorem 39 ([110]). *Let $q = q_\star(t)$ be a solution of (10.5) with $u = u_\star(t)$, which are of class C^1 on $[0, T]$. Suppose $f_1(\phi), \ldots, f_n(\phi)$ are functions of class C^2 that represent an alternative parametrization (10.7) of this motion such that the smooth matrix functions $L(y, \phi)$ and $N(y, \phi)$, given by (10.23) and, respectively, by (10.29), are invertible in a vicinity (10.4) of the orbit. Then relations (10.20) and (10.21) define the generalized coordinates $\phi, y_1, \ldots, y_{n-1}$ for the mechanical system (10.1). Moreover, in the new generalized coordinates, the dynamics (10.5) are governed by Eqs. (10.28), (10.34), where the former equation can be simplified, via the coordinate-feedback transformation (10.30), to the form (10.31).*

10.1 Virtual Constraint Approach

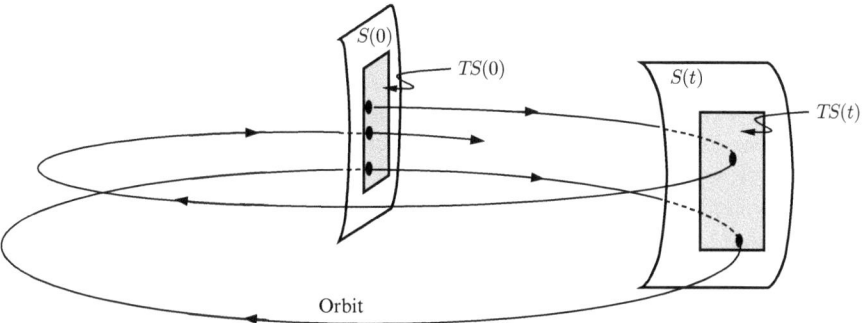

Fig. 10.1 Moving Poincaré section for the periodic trajectory $(\phi_\star, \dot\phi_\star)$, where $TS(\cdot)$ denotes the tangent space

10.1.4 Moving Poincaré Section, Transverse Coordinates, and Transverse Linearization

A linearization of an autonomous system has long been recognized [6] not to be asymptotically stable [6]; for instance, the homogeneous version (10.10) of system (10.34) under certain conditions generates cycles, thereby yielding the linearization of (10.10) along such a cycle asymptotically unstable. This motivation has brought into play new state variables decomposing the system dynamics into a scalar variable ϕ, which represents the position along a target orbit, and the remaining coordinates, which represent the dynamics transverse to the orbit. Since the local properties of the system dynamics around the orbit are independent of ϕ, this variable can be safely disregarded beyond the orbit, whereas the transverse coordinate defines a moving Poincaré section [77].

A family of $(2n-1)$-dimensional C^1-smooth surfaces $\{S(t), t \in [0, T]\}$ of class C^1 (see Fig. 10.1) is said to be a *moving Poincaré section* associated with the solution $q_\star(t), t \in [0, T]$, if

- Surfaces $S(\cdot)$ are locally disjoint; that is, there exists a $\varepsilon > 0$ such that $S(\tau_1) \cap S(\tau_2) \cap \mathcal{O}_\varepsilon = \emptyset$, for all $\tau_1 \in [0, T), \tau_2 \in (0, T], \tau_1 \neq \tau_2$.
- Each of the surfaces $S(\cdot)$ locally intersects the orbit only in one point; that is, there exists $\varepsilon > 0$ such that $S(\tau) \cap \{(q_\star(t), \dot q_\star(t)), |t - \tau| < \varepsilon\} \cap \mathcal{O}_\varepsilon(q_\star) = \{(q_\star(\tau), \dot q_\star(\tau))\}$ for each $\tau \in [0, T]$.
- The surfaces $S(\cdot)$ are smoothly parameterized by time; that is, there exists $\xi_s \in C^1(\mathbb{R}^{2n+1})$ such that $S(t) \cap \mathcal{O}_\varepsilon(q_\star) = \{(q, \dot q) \in \mathbb{R}^{2n} : \xi_s(q, \dot q, t) = 0\} \cap \mathcal{O}_\varepsilon(q_\star)$.
- The surfaces $S(\cdot)$ are transversal to the orbit (10.3); that is, for all $t \in [0, T]$,

$$\left[\frac{\partial \xi_s}{\partial q}\bigg|_{\substack{q = q_\star(t) \\ \dot q = \dot q_\star(t)}}\right]^T \dot q_\star(t) + \left[\frac{\partial \xi_s}{\partial \dot q}\bigg|_{\substack{q = q_\star(t) \\ \dot q = \dot q_\star(t)}}\right]^T \ddot q_\star(t) \neq 0.$$

Given a moving Poincaré section $\{S(t), t \in [0, T]\}$ associated with the motion $q_\star(t)$, the state coordinates $(q, \dot{q})$ of (10.1) can be (locally) changed into $(\phi, x_\perp)$, where the scalar variable $\phi(t)$ parameterizes position along the curve (trajectory) defined by $q_\star(t)$ and the $(2n-1)$-dimensional vector $x_\perp(t)$ defines location on the surfaces $S(t)$. The components of the vector $x_\perp$ are recognized as *transverse coordinates*, while the $x_\perp$-dynamics are called *transverse*.

The dynamics (10.1), rewritten in $(\phi, x_\perp)$-coordinates and linearized along the solution $q_\star(t)$, $t \in [0, T]$, give rise to the linear time-varying control system of dimension $2n$ defined on $t \in [0, T]$. The $(2n-1)$-dimensional subsystem that corresponds to linearization of the dynamics of the transverse coordinates $x_\perp$ is referred to as *transverse linearization*.

If the conditions of Theorem 39 hold, then the dynamics of the mechanical system (10.5) can be rewritten in the form (10.28), (10.34). It turns out that the system (10.28), (10.34) possesses a natural choice of $(2n-1)$-transverse coordinates

$$x_\perp = [I(\phi_\star(0), \dot{\phi}_\star(0), \phi, \dot{\phi}), y, \dot{y}]^T, \qquad (10.35)$$

which can be introduced in a vicinity of the solution

$$y_1 = y_{\star_1} \equiv 0, \ldots, y_{n-1} = v_{\star_{n-1}} \equiv 0, \qquad \phi = \phi_\star \qquad (10.36)$$

of (10.5), and for which the transverse linearization is computed analytically (see [112] for details). The scalar transverse coordinate $I(\phi_\star(0), \dot{\phi}_\star(0), \phi, \dot{\phi})$, given by (10.18), can serve as a measure of the distance to the trajectory $q_\star(t)$, and when viewed along a solution $(\phi(t), \dot{\phi}(t))$ of (10.33), its time derivative

$$\frac{d}{dt} I = \dot{\phi} \left[\frac{2}{\alpha(\phi)} g - \frac{2\beta(\phi)}{\alpha(\phi)} I \right] \qquad (10.37)$$

is calculated according to [112].

10.2 Orbital Synthesis Procedure via $\mathcal{H}_\infty$-Position Feedback

Applying the virtual constraint approach yields a constructive framework for orbitally stabilizing position-feedback synthesis around a prespecified virtual constraint (10.7). It is assumed that the virtual constraint (10.7) is chosen in such a manner that the virtual limit system (10.10) possesses a T-periodic solution $\phi_\star(t) = \phi_\star(t + T)$ for all $t \geq 0$ (e.g., it is the case if the conditions of Theorem 38 are satisfied). Then the corresponding dynamics

$$q_{\star_1}(t) = f_1(\phi_\star(t)), \ldots, q_{\star_n}(t) = f_n(\phi_\star(t)) \qquad (10.38)$$

of the underactuated mechanical system (10.5), properly initialized and driven by an appropriate feedback $u_\star$ along the virtual constraint (10.7), would also be T-periodic. The existence of the periodic orbit $q_\star(t) = (q_{\star_1}(t), \ldots, q_{\star_n}(t))^T$ as well as the existence of the nominal position feedback $u_\star$ are thus postulated a priori.

The $\mathcal{H}_\infty$-periodic-motion generation to be developed via position feedback aims to asymptotically stabilize the closed-loop system (10.5) around the orbit $q_\star(t)$ thus specified while also attenuating external disturbances potentially affecting the system.

The proposed position-feedback $\mathcal{H}_\infty$ synthesis, orbitally stabilizing the underactuated mechanical system (10.5) around the chosen orbit $\phi_\star(t)$, is as follows.

- In a vicinity of the periodic solution $q_\star(t)$, system (10.5) is represented in the $(\phi, x_\perp)$-coordinates, where the scalar variable $\phi(t)$ parameterizes position along $q_\star(t)$ and the $(2n-1)$-dimensional transverse vector (10.35) defines location on the moving Poincaré section $S(t)$.
- Provided that the matrices $L(y, \phi)$ and $N(y, \phi)$, given by (10.23) and (10.29), respectively, are invertible, the coordinate-feedback transform (10.30) allows one to bring the original system (10.5) into the form (10.31), (10.34), (10.37).
- The position-feedback $\mathcal{H}_\infty$ synthesis of Sect. 6.2, specified for the stabilization of the transformed system (10.31), (10.34), (10.37) around the periodic solution (10.36), solves the periodic-motion-generation problem in question.

Alternatively, the above procedure can be simplified by applying the $\mathcal{H}_\infty$ synthesis of Sect. 6.2 to the $(2n-1)$-dimensional transverse dynamics (10.31), (10.37) only. Such a reduced-order synthesis becomes possible (cf. [112, Theorem 3]) due to the property, established by Lemma 12, of the scalar component (10.18) of the transverse vector $[I(\phi_\star(0), \dot{\phi}_\star(0), \phi, \dot{\phi}), y, \dot{y}]^T$ to preserve the zero value along the prespecified periodic solution $\phi_\star(t)$. Capabilities of the proposed periodic motion generation are subsequently illustrated in a numerical study of a laboratory test bed having 3 DOF and one degree of underactuation.

10.3 Case Study: A 3-DOF Underactuated Helicopter

The analysis and synthesis of helicopter models have attracted much attention among researchers and have been presented in numerous publications, such as [11, 41, 69, 82] (see also references therein), to name a few. A helicopter prototype, manufactured by the Quanser Company, is presently used to illustrate the proposed $\mathcal{H}_\infty$-periodic-motion-generation procedure. It is a rigid body with a spherical joint at the suspension point. Rotation of the prototype is permitted around the suspension point in any direction. There are two propellers, which are symmetrically attached at the end of the body and which can be actuated individually. A counterweight is installed at the other end so that the gravitational forces become negligible. The prototype is depicted in Fig. 10.2; it represents an underactuated 3-DOF mechanical system, actuated by two motors.

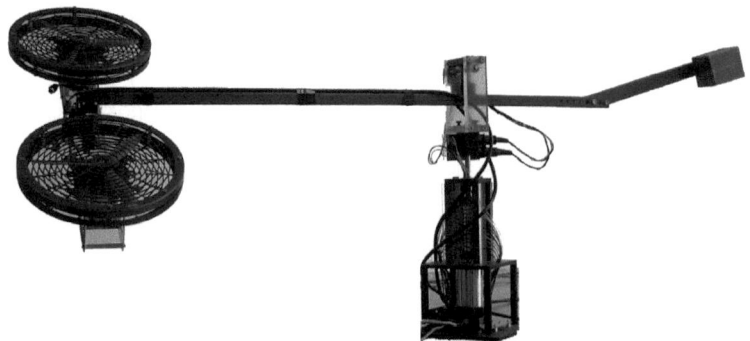

Fig. 10.2 3-DOF helicopter prototype

10.3.1 Dynamic Model

The dynamics of the 3-DOF helicopter prototype, used throughout, are derived from the Newton–Euler equations (see [115])

$$J\dot{\omega} + \omega \times (J\omega) = \tau, \qquad (10.39)$$

where $\omega = [\omega_x, \omega_y, \omega_z]^T$ is the angular velocity, $J = diag(J_x, J_y, J_z)$ is the inertia matrix of the rigid body attached at the suspension point about which it can rotate. The vector of torques applied can be reduced to $\tau = [\tau_x, \tau_y, 0]^T$ considering hereby the one degree of underactuation with respect to the rotation axis of the mechanism.

Similar to [130], the above model has been simplified in two ways. Firstly, the gravity is neglected by assuming that the body is statically balanced about the suspension point; that is, the counterweight is chosen to compensate for the body gravitation, thereby yielding the center of mass at the pivot point. Secondly, air resistance and friction forces are not taken into account. These simplifications are made to facilitate the exposition, thus focusing on intrinsic features of the virtual constraint approach to solving the periodic-motion-planning problem. Augmenting the above model with gravitational, air resistance, and frictional forces is indeed possible by following, for instance, the method proposed in [28].

According to the Newton–Euler formulation, all the variables are expressed in the body-attached frame, shown in Fig. 10.3, along with the generalized coordinates in the inertial frame. To obtain the angular velocity of the body expressed in the inertial frame, one only needs to compute a rotation matrix that transforms coordinates from one frame to the next. For this purpose, let's introduce the vector

$$q = [\phi, \theta, \psi]^T \qquad (10.40)$$

of the $Z - Y - X$ Euler angles, which are also referred to as yaw, pitch, and roll, respectively. This vector corresponds to the generalized coordinates describing the

10.3 Case Study: A 3-DOF Underactuated Helicopter

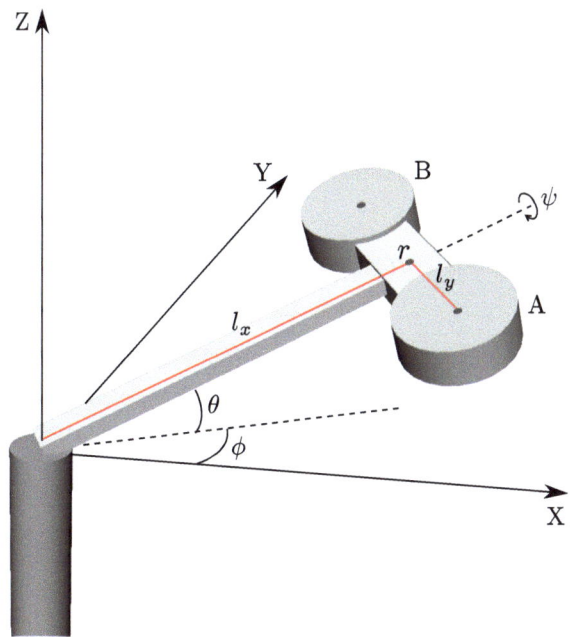

Fig. 10.3 Graphical representation of the generalized coordinates of the 3-DOF helicopter prototype

orientation of the rigid body. Then, the angular velocity ω, expressed in the body frame, relates to the rates of the change of the generalized coordinate vector (10.40) as follows [130]:

$$\omega = A_{zyx}\dot{q} = \begin{bmatrix} 0 & -\sin(\phi) & \cos(\phi)\cos(\theta) \\ 0 & \cos(\phi) & \sin(\phi)\cos(\theta) \\ 1 & 0 & -\sin(\theta) \end{bmatrix} \begin{bmatrix} \dot{\phi} \\ \dot{\theta} \\ \dot{\psi} \end{bmatrix}, \quad (10.41)$$

where $\dot{q} = [\dot{\phi}, \dot{\theta}, \dot{\psi}]^T$ and

$$A_{zyx} = \begin{bmatrix} 0 & -\sin(\phi) & \cos(\phi)\cos(\theta) \\ 0 & \cos(\phi) & \sin(\phi)\cos(\theta) \\ 1 & 0 & -\sin(\theta) \end{bmatrix}. \quad (10.42)$$

System (10.39)–(10.42) can be represented in the inertial framework by the differential equation

$$M(q)\ddot{q} + \vartheta(q,\dot{q}) = Bu, \quad (10.43)$$

where

$$u = [\tau_x, \tau_y]^T, \ B = \begin{bmatrix} 1 & 0 \\ 0 & 1 \\ 0 & 0 \end{bmatrix}, \ M(q) = JA_{zyx}, \ \vartheta(q,\dot{q}) = J\dot{A}_{zyx}\dot{q} + \omega \times (J\omega). \quad (10.44)$$

The latter two matrices admit the explicit computation

$$M(q) = \begin{bmatrix} 0 & -J_x s_\phi & J_x c_\phi c_\theta \\ 0 & J_y c_\phi & J_y s_\phi c_\theta \\ J_z & 0 & -J_z s_\theta \end{bmatrix}, \qquad (10.45)$$

$$\vartheta(q,\dot{q}) = \begin{bmatrix} -J_x c_\phi \dot{\phi}\dot{\theta} - J_x(c_\phi s_\theta \dot{\theta} + s_\phi c_\theta \dot{\phi})\dot{\psi} \\ -J_y s_\phi \dot{\phi}\dot{\theta} + J_y(c_\phi c_\theta \dot{\phi} - s_\phi s_\theta \dot{\theta})\dot{\psi} \\ -J_z c_\theta \dot{\theta}\dot{\psi} \end{bmatrix}$$

$$+ \begin{bmatrix} (c_\phi \dot{\theta} + s_\phi c_\theta \dot{\psi}) J_z(\dot{\phi} - s_\theta \dot{\psi}) \\ -(-s_\phi \dot{\theta} + c_\phi c_\theta \dot{\psi}) J_z(\dot{\phi} - s_\theta \dot{\psi}) \\ (-s_\phi \dot{\theta} + c_\phi c_\theta \dot{\psi}) J_y(c_\phi \dot{\theta} + s_\phi c_\theta \dot{\psi}) \end{bmatrix}$$

$$+ \begin{bmatrix} -(\dot{\phi} - s_\theta \dot{\psi}) J_y(c_\phi \dot{\theta} + s_\phi c_\theta \dot{\psi}) \\ (\dot{\phi} - s_\theta \dot{\psi}) J_x(-s_\phi \dot{\theta} + c_\phi c_\theta \dot{\psi}) \\ -(c_\phi \dot{\theta} + s_\phi c_\theta \dot{\psi}) J_x(-s_\phi \dot{\theta} + c_\phi c_\theta \dot{\psi}) \end{bmatrix}, \qquad (10.46)$$

where the notations $c_x = \cos(x)$ and $s_x = \sin(x)$ have been adopted for ease of reference.

While being represented in the inertial framework, the state equation (10.43) is no more than the Lagrangian formulation (10.5) of the helicopter prototype dynamics planning, which can now be attained by applying the virtual constraint approach.

10.3.2 The Motion Planning

For the periodic-motion planning of the 3-DOF rigid body (10.43)–(10.46) about the yaw axis, we've chosen the coordinate ϕ as an independent variable to be used for parameterizing such a motion. The other two generalized coordinates are then geometrically related to ϕ to predetermine a path in the configuration space. Hence, the virtual holonomic constraint takes the form

$$q = \begin{bmatrix} \phi \\ \theta \\ \psi \end{bmatrix} := F(\phi) = \begin{bmatrix} \phi \\ f_1(\phi) \\ f_2(\phi) \end{bmatrix}. \qquad (10.47)$$

Provided that for the underactuated system (10.43)–(10.46), there exists a control input $u = u_\star(t)$ that makes the virtual holonomic constraint (10.47) invariant, then the overall closed-loop system can be represented by the reduced-order dynamics (10.10) of the form

$$\alpha(\phi)\ddot{\phi} + \beta(\phi)\dot{\phi}^2 + \gamma(\phi)\phi = 0. \qquad (10.48)$$

The solutions of this virtually constrained system define feasible motions of the body with precise synchronization given by (10.47). It means that the whole motion is parameterized by the evolution of the chosen configuration variable ϕ. The smooth functions $\alpha(\phi)$, $\beta(\phi)$, and $\gamma(\phi)$, given by (10.11), are now specified as

$$\alpha(\phi) = J_z[1 - f_2'(\phi)s_{f_1(\phi)}],$$
$$\beta(\phi) = -J_z s_{f_1(\phi)} f_2''(\phi)$$
$$+ \begin{bmatrix} s_\phi(J_x - J_y)c_\phi \\ -c_\phi s_\phi(J_x - J_y)c_{f_1(\phi)}^2 \end{bmatrix}^T \begin{bmatrix} f_1'(\phi) \\ f_2'(\phi) \end{bmatrix}^2 \qquad (10.49)$$
$$- [(J_x - J_y)c_\phi^2 + (J_y - J_x)s_\phi^2 + J_z]c_{f_1(\phi)} f_1'(\phi) f_2'(\phi),$$
$$\gamma(\phi) = 0.$$

Recall that the virtual limit system (10.48) is integrable once $\alpha(\phi) \neq 0$ on the interval of interest. Specifically, the integral function (10.18) preserves its zero value along the solution of (10.48), initiated at $(\phi(0), \dot\phi(0)) = (\phi_0, \dot\phi_0)$.

Eventually, one can analytically compute the nominal control torque

$$\tau_\star = \begin{bmatrix} 1 & 0 & 0 \\ 0 & 1 & 0 \end{bmatrix} [J\dot\omega + \omega \times (J\omega)] \Big|_{\substack{q = F(\phi_\star) \\ \dot q = F'(\phi_\star)\dot\phi_\star \\ \ddot q = F''(\phi_\star)\dot\phi_\star^2 + F'(\phi_\star)\ddot\phi_\star}} \qquad (10.50)$$

which generates the desired solution $\phi = \phi_\star(t)$ of (10.48) along the virtual holonomic constraint (10.47).

10.3.3 $\mathcal{H}_\infty$ Orbitally Stabilizing Synthesis

The control objective for the 3-DOF underactuated helicopter prototype (10.43)–(10.46) to generate a periodic motion is to design a nonlinear $\mathcal{H}_\infty$-position-feedback controller that makes a prespecified T-periodic motion

$$q_\star(t) = [\phi_\star(t), f_1(\phi_\star(t)), f_2(\phi_\star(t))]^T \qquad (10.51)$$

along a holonomic virtual constraint (10.47), generated by the feasible torque (10.50), to be asymptotically stable. Along with this, the effect of an external disturbance $w_1 \in \mathbb{R}^3$ on the perturbed model

$$M(q)\ddot q + \vartheta(q, \dot q) = Bu + w_1 \qquad (10.52)$$

and that of a position measurement noise $w_0 \in \mathbb{R}^3$ should also be attenuated.

Note that according to the state equations (10.43)–(10.46), the desired periodic motion (10.51) is governed by

$$M(q_\star)\ddot{q}_\star + \vartheta(q_\star,\dot{q}_\star) = Bu_\star, \qquad (10.53)$$

where $u_\star = (\tau_{\star_x},\tau_{\star_y})^T$ is composed of the entries of the feasible torque $\tau_\star = (\tau_{\star_x},\tau_{\star_y},0)^T$.

In order to formally state the $\mathcal{H}_\infty$-periodic-motion-generation problem, let's decompose the controller to be designed into the form

$$u = u_\star + v, \qquad (10.54)$$

consisting of the trajectory feedforward compensator $u_\star(t)$, governed by (10.53), and a local disturbance attenuator $v(t)$, internally stabilizing the closed-loop system locally around the desired trajectory. Let's then represent the state equations (10.43)–(10.46) in terms of the state-deviation vector $x = (x_1,x_2)^T$, the first component of which is the position deviation

$$x_1 = \begin{bmatrix} \phi - \phi_\star, & \theta - f_1(\phi_\star), & \psi - f_2(\phi_\star) \end{bmatrix}^T \qquad (10.55)$$

from the desired trajectory (10.51), and the second component of which is the velocity deviation

$$x_2 = \begin{bmatrix} \dot{\phi} - \dot{\phi}_\star, & \dot{\theta} - \dot{f}_1(\phi_\star), & \dot{\psi} - \dot{f}_2(\phi_\star) \end{bmatrix}^T \qquad (10.56)$$

from the desired velocity $\dot{q}_\star = [\dot{\phi}_\star, \dot{f}_1(\phi_\star), \dot{f}_2(\phi_\star)]^T$. Now, taking the control decomposition (10.54) into account and subtracting (10.53) from (10.52) yield the state-deviation equations

$$\dot{x}_1 = x_2,$$
$$\dot{x}_2 = M^{-1}(x_1 + q_\star)[\vartheta(q_\star,\dot{q}_\star) - \vartheta(x_1 + q_\star, x_2 + \dot{q}_d)$$
$$+ M(q_\star)\ddot{q}_\star + Bv + w_1] - \ddot{q}_\star, \qquad (10.57)$$

$$z = \rho \begin{bmatrix} 0_{2\times 1} \\ x_1 \end{bmatrix} + \begin{bmatrix} I_2 \\ 0_{3\times 2} \end{bmatrix} v(t), \qquad (10.58)$$

$$y = x_1 + w_0, \qquad (10.59)$$

where $z \in \mathbb{R}^5$ is the output to be controlled, $y \in \mathbb{R}^3$ is the available position measurement, $w = (w_0^T, w_1^T)^T \in \mathbb{R}^6$ is the disturbance to be attenuated, and the weight parameter $\rho \in \mathbb{R}$ is positive.

The periodic-motion-generation problem in question can thus be stated as a standard $\mathcal{H}_\infty$-control problem for the error system (10.57)–(10.59). Let a desired T-periodic motion (10.51) for the helicopter model (10.43)–(10.46) to track be

10.3 Case Study: A 3-DOF Underactuated Helicopter

generated by the feasible torque (10.50) along a holonomic virtual constraint (10.47) so that the virtual limit system (10.48) possesses an underlying T-periodic solution $\phi_\star(t)$, resulting in (10.51). Given a real number $\gamma > 0$, it is required to find (if any) a causal dynamic feedback controller (10.54), consisting of the trajectory feedforward compensator $u_\star(t)$, governed by (10.53), and a local disturbance attenuator $v = \mathcal{K}(y, t)$ with internal state $\xi \in \mathbb{R}^s$ such that the undisturbed error system (10.57)–(10.59) in the closed loop is uniformly asymptotically stable and its $\mathcal{L}_2$ gain is locally less than γ: That is, inequality

$$\int_{t_0}^{t_1} \|z(t)\|^2 dt < \gamma^2 \int_{t_0}^{t_1} \|w(t)\|^2 dt \tag{10.60}$$

is satisfied for all $t_1 > t_0$ and all piecewise-continuous functions $w(t) = (w_0^T(t), w_1^T(t))^T$ for which the trajectory $(x^T(t), \xi^T(t))^T$ of the closed-loop system, starting from the origin, remain in some neighborhood of the origin for all $t \in [t_0, t_1]$.

Since, by an appropriate choice of the functions f_1 and f_2 constituting the virtual constraint (10.47), the periodic trajectory $q_\star(t)$ proves to be of class C^2, whereas the right-hand side of the error dynamics (10.57)–(10.59) turns out to be twice continuously differentiable in x for all t, the above $\mathcal{H}_\infty$-control problem can be resolved within the general approach, developed in Chap. 6 for a generic system (6.1)–(6.3), which is now specified as follows:

$$f(x, t) = \begin{bmatrix} x_2 \\ M^{-1}(x_1 + q_\star)[\vartheta(q_\star, \dot{q}_\star) - \vartheta(x_1 + q_\star, x_2 + \dot{q}_\star)] \end{bmatrix}$$
$$+ \begin{bmatrix} 0 \\ M^{-1}(x_1 + q_\star) M(q_d) \ddot{q}_\star - \ddot{q}_\star \end{bmatrix}, \tag{10.61}$$

$$g_1(x, t) = \begin{bmatrix} 0_{3\times 3} & 0_{3\times 3} \\ 0_{3\times 3} & M^{-1}(x_1 + q_\star) \end{bmatrix},$$

$$g_2(x, t) = \begin{bmatrix} 0_{3\times 2} \\ M^{-1}(x_1 + q_\star) B \end{bmatrix},$$

$$h_1(x, t) = \rho \begin{bmatrix} 0_{2\times 1} \\ x_1 \end{bmatrix}, \quad h_2(x) = x_1,$$

$$k_{12}(x) = \begin{bmatrix} I_2 \\ 0_{3\times 2} \end{bmatrix}, \quad k_{21}(x) = \begin{bmatrix} I_3 & 0_{3\times 3} \end{bmatrix}. \tag{10.62}$$

Theorem 40. *Let the virtual constraint (10.47) be chosen in such a manner that while being properly initialized on and driven along this constraint, the helicopter model (10.43)–(10.46) possesses a twice continuously differentiable T-periodic solution (10.51). Apart from this, let Conditions $C1'$, $C2'$ of Sect. 6.2*

Table 10.1 Parameter values of the experimental 3-DOF helicopter

Parameter	Value	Units
Inertias	$J_x = 0.0364$	kg m^2
	$J_y = 0.91$	kg m^2
	$J_z = 0.91$	kg m^2
Lengths	$l_x = 0.66$	m
	$l_y = 0.177$	m

hold for the periodic matrix functions $A(t)$, $B_1(t)$, $B_2(t)$, $C_1(t)$, $C_2(t)$, governed by (6.35), (10.61), (10.62), and let $(P_\varepsilon(t), Z_\varepsilon(t))$ denote the corresponding T-periodic positive definite solution of (6.40), (6.41) under some $\varepsilon > 0$. Then the dynamic position feedback (10.54), consisting of the trajectory feedforward compensator $u_\star(t)$, governed by (10.53), and a disturbance attenuator

$$\dot{\xi} = f(\xi, t) + \left[\frac{1}{\gamma^2} g_1(\xi, t) g_1^T(\xi, t) - g_2(\xi, t) g_2^T(\xi, t) \right] P_\varepsilon(t) \xi$$
$$+ Z_\varepsilon(t) C_2^T(t) [y(t) - h_2(\xi, t)],$$
$$v = -g_2^T(\xi, t) P_\varepsilon(t) \xi, \tag{10.63}$$

subject to (10.61), (10.62), is a local solution of the $\mathcal{H}_\infty$-periodic-motion-generation problem for the helicopter prototype (10.43)–(10.46).

Proof. By applying Theorem 25 to the error system (10.57)–(10.59), we may verify the validity of Theorem 40.

10.3.4 Simulation Results

In order to illustrate the capabilities of the developed approach, we undertook a numerical study for the laboratory helicopter prototype with different periodic motions to follow. The physical parameters of the laboratory 3-DOF helicopter are shown in Fig. 10.3 and were extracted from the prototype manual [99]; they are listed in Table 10.1.

10.3.4.1 Periodic-Motion Generation with Constant Pitch and Constant Roll

If we specify the virtual holonomic constraint (10.47) with

$$f_1(\phi) = \kappa_1,$$
$$f_2(\phi) = \kappa_2, \tag{10.64}$$

10.3 Case Study: A 3-DOF Underactuated Helicopter

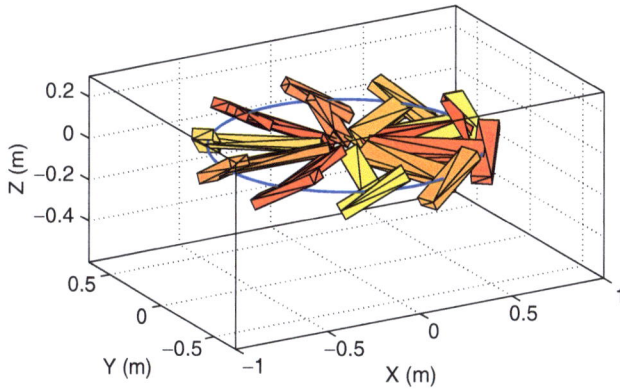

Fig. 10.4 Rigid body motion for constant roll and constant pitch (10.64)

the generated periodic motion exhibits the constant pitch $\theta = \kappa_1$ and constant roll $\psi = \kappa_2$. Figure 10.4 shows the configuration of the 3-DOF helicopter model in inertial frame at certain equally distanced time instants where the trajectory followed by the endpoint of the body is additionally depicted.

Substituting the desired virtual constraint functions (10.64) into (10.49) yields the dynamics of the virtual limit system (10.48) in the form

$$J_z \ddot{\phi} = 0, \qquad (10.65)$$

which ensures a constant angular velocity $\dot{\phi}_\star$ along the resulting periodic motion $\phi_\star(t) = \dot{\phi}_\star t + \phi(0)$. Furthermore, one can specify a desired value of the angular velocity.

In the simulation runs, the virtual constraint parameters are set to $\kappa_1 = 0$ and $\kappa_2 = 0.5$, and the periodic motion to be tracked along the path (10.64) is selected according to

$$\phi_\star(t) \in [0, 2\pi] \text{ rad}$$
$$\text{for } t \in [0, T] \text{ with } T = 12.57 \text{ s, and} \qquad (10.66)$$
$$(\phi_\star(0), \dot{\phi}_\star(0)) = (0 \text{ rad}, 0.5 \text{ rad/s}),$$

where initial conditions have been prespecified in such a manner that the period T of the motion verifies the conditions $\phi_\star(T) = 2\pi$ and $\dot{\phi}_\star(T) = 0.5$ for the cyclic yaw variable $\phi(t)$ and for its angular velocity $\dot{\phi}(t)$, respectively.

The orbitally stabilizing synthesis around the periodic motion thus specified is achieved by implementing the nonlinear $\mathcal{H}_\infty$ controller (10.63), properly tuned with the weight parameter $\rho = 2$, $\gamma = 260$, and $\varepsilon = 1$. The simulation results, presented in Fig. 10.5, match the initial conditions $\phi_1(0) = 0$, $\phi_2(0) = 0.5$, $\theta_1(0) = 0.25$, $\theta_2(0) = 0$, $\psi_1(0) = 0$, $\psi_2(0) = 0$, the vector harmonic disturbance $w_1 = (w_{11}, w_{12}, w_{13})^T$, whose components

$$w_{11}(t) = 0.1 \cos(50t), \ w_{12}(t) = 0.2 \cos(70t), \ w_{13}(t) = 0.3 \cos(70t), \quad (10.67)$$

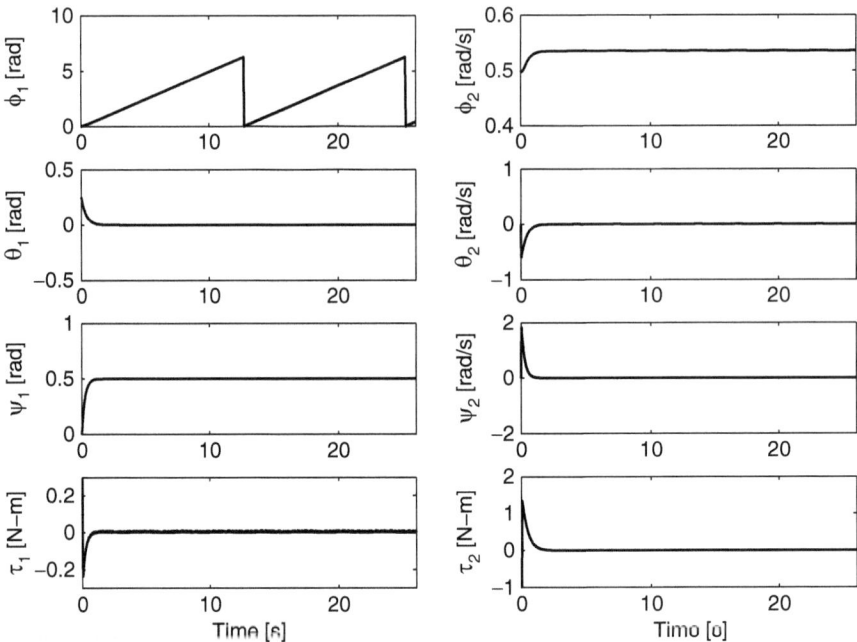

Fig. 10.5 Orbitally stabilizing synthesis with constant pitch and constant roll under harmonic disturbances

affect the yaw, pitch, and roll dynamics, respectively, and the measurement noise $w_0 = (w_{01}, w_{02}, w_{03})^T$ with the components

$$w_{0i} = 0.001\cos(2t), \quad i = 1, 2, 3. \tag{10.68}$$

A good performance of the closed-loop system is concluded from Fig. 10.5, where the proposed desired behavior (10.64)–(10.66) is approached, while the effect of the external disturbance is attenuated. The torques applied to each actuator are also depicted in Fig. 10.5.

10.3.4.2 Periodic-Motion Generation with Cyclic Pitch and Constant Roll

When we specify the virtual holonomic constraint (10.47) with

$$\begin{aligned} f_1(\phi) &= \kappa_1 \cos(n\phi), \\ f_2(\phi) &= \kappa_2, \end{aligned} \tag{10.69}$$

the generated periodic motion exhibits the cyclic pitch $\theta = \kappa_1 \cos(n\phi)$ and constant roll $\psi = \kappa_2$.

10.3 Case Study: A 3-DOF Underactuated Helicopter

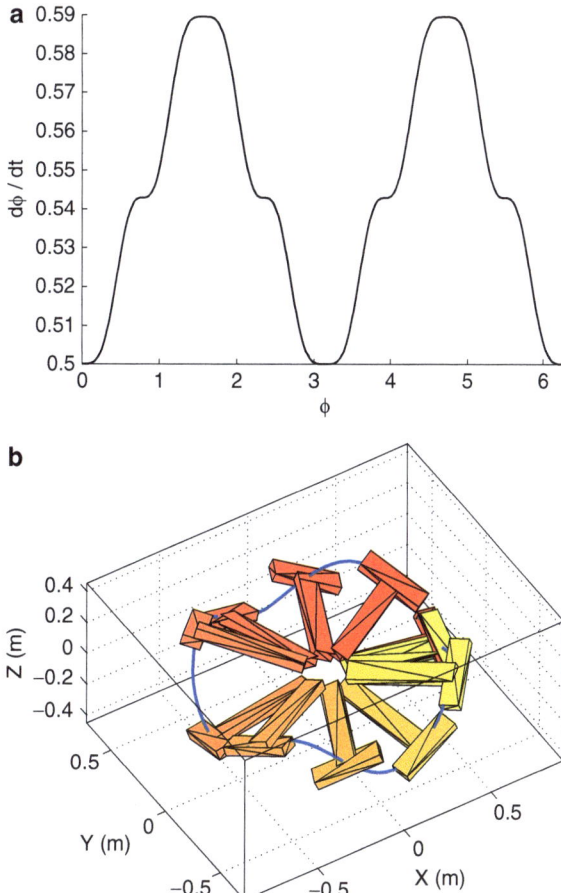

Fig. 10.6 (a) Trajectory (10.71) on the phase portrait for (10.70) and (b) rigid body motion for cyclic pitch and constant roll (10.69)

In the simulation runs, the virtual constraint parameters are set to $\kappa_1 = 0.2$, $\kappa_2 = 0.4$, and $n = 4$. If we substitute the desired virtual constraint functions (10.69) into (10.49), we obtain the dynamics of the virtual limit system (10.48) in the form

$$J_z \ddot{\phi} + [16\kappa_1^2 \sin^2(4\phi) \sin(\phi)(J_x - J_y) \cos(\phi)]\dot{\phi}^2 = 0. \tag{10.70}$$

For the purpose of the periodic-motion generation, the following periodic trajectory

$$\begin{aligned} &\phi_\star(t) \in [0, 2\pi] \text{ rad} \\ &\text{for } t \in [0, T] \text{ with } T = 11.60 \text{ s, and} \\ &(\phi_\star(0), \dot{\phi}_\star(0)) = (0 \text{ rad}, 0.5 \text{ rad/s}) \end{aligned} \tag{10.71}$$

is selected along the path given by (10.69). The initial conditions in (10.71) have been prespecified in such a manner that the period T of the rotation verifies the conditions $\phi_\star(T) = 2\pi$ and $\dot{\phi}_\star(T) = 0.5$ for the cyclic yaw variable $\phi(t)$ and for

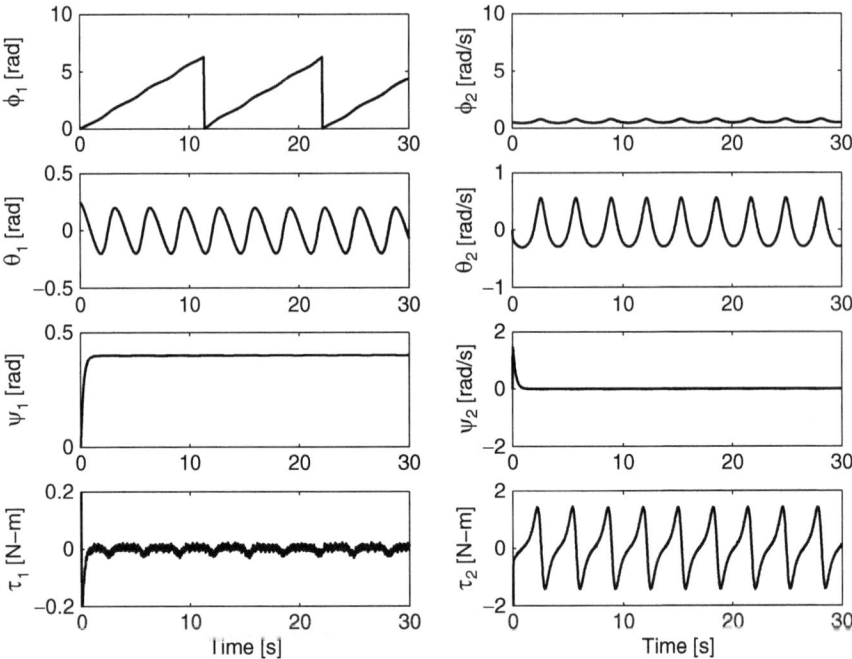

Fig. 10.7 Orbitally stabilizing synthesis with cyclic pitch and constant roll under harmonic disturbances

its angular velocity $\dot{\phi}(t)$, respectively. The resulting body motion (10.69)–(10.71), depicted in Fig. 10.6, corresponds to the solution of the reduced dynamics evolving within constraint (10.69).

The orbitally stabilizing synthesis around the periodic motion thus specified is achieved by implementing the nonlinear $\mathcal{H}_\infty$ controller (10.63), properly tuned with the weight parameter $\rho = 1$, $\gamma = 220$, and $\varepsilon = 1$. The simulation results, presented in Fig. 10.7, match the initial conditions $\phi_1(0) = 0, \phi_2(0) = 0.5, \theta_1(0) = 0.25, \theta_2(0) = 0, \psi_1(0) = 0 = \psi_2(0) = 0$, the external harmonic disturbance (10.67), affecting the body motion, and the measurement noise (10.68). A good performance of the closed-loop system is concluded from Fig. 10.7, where the proposed desired behavior (10.69)–(10.71) is evidently achieved while the effect of the external disturbance is attenuated. The torques applied to each actuator are also depicted in Fig. 10.7.

Chapter 11
LMI-Based $\mathcal{H}_\infty$ Synthesis of the Current Profile in Tokamak Plasmas

The $\mathcal{H}_\infty$ stabilization of the plasma current profile in a toroidal chamber with magnetic coils (tokamak [129]), recently proposed in [54], is invoked to illustrate the LMI extension to the PDE setting. The subsequent control design is based on a one-dimensional (1) resistive diffusion equation of the magnetic flux that governs the plasma current profile evolution. While being of a parabolic type, the underlying PDE is not, however, similar to that of Chap. 7 as it contains a non-self-adjoint infinitesimal operator in the state equation, a feature that is not typical in the existing literature on DPS control. Thus, the LMI $\mathcal{H}_\infty$ synthesis of Chap. 7 is to be extended to a specific parabolic PDE with a non-self-adjoint infinitesimal operator.

The proposed distributed control is a proportional-integral state feedback that takes into account both interior and boundary engineering actuators. A target profile, which should constitute the steady state of the closed-loop system, is designed a priori using manipulatable system inputs such as the loop voltage, the lower hybrid power, and the wave refractive index. A model-based optimization procedure is then applied at the simulation stage to derive the engineering plant inputs related to both inductive and noninductive current drive means. Numerical simulations, extracted from [54], were performed using typical Tore Supra values [107] and yielded quite positive results with promising robustness properties.

11.1 Modeling and Problem Statement

Under standard assumptions such as axisymmetry, magnetohydrodynamics equilibrium, averaging over the magnetic surfaces, and cylindrical approximation (see, e.g., [21, 134] and Fig. 11.1), the evolution of the plasma current profile q, which is the safety factor to be controlled, can be obtained by solving the following 1D PDE:

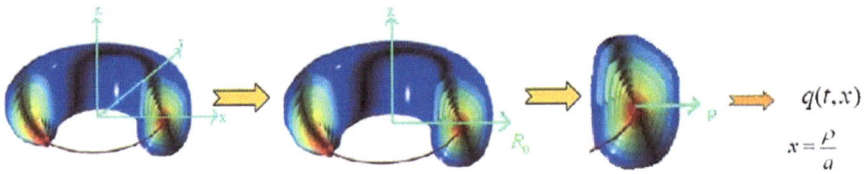

Fig. 11.1 One-dimensional geometry-simplified formulation

$$\begin{cases} \frac{\partial \psi}{\partial t}(x,t) = \frac{\eta_{\|}(x,t)}{\mu_0 a^2} \frac{1}{x} \frac{\partial}{\partial x}\left(x \frac{\partial \psi}{\partial x}(x,t)\right) \\ \qquad + \eta_{\|}(x,t) R_0 j_{ni}(x,t); \\ \frac{\partial \psi}{\partial x}(x,t)\Big|_{x=0} = 0, \quad \frac{\partial \psi}{\partial t}(t,1) = -V_0(t), \end{cases} \quad (11.1)$$

and determining it according to

$$q(x,t) = -\frac{a^2 x B_0}{\frac{\partial \psi}{\partial x}(x,t)}. \quad (11.2)$$

In the above relations, $x \in [0,1]$ is the 1D (radial) profile coordinate, t is the time, $\psi(x,t)$ the magnetic flux, R_0 and a are, respectively, the major radius and the minor radius of the plasma boundary (both are assumed to be fixed), $\mu_0 > 0$ is the permeability of vacuum, $\eta_{\|}(x,t)$ is the parallel electrical resistivity of the plasma, $V_0(t)$ is the plasma loop voltage, B_0 is the toroidal magnetic field at R_0, and $j_{ni}(x,t)$ is the noninductive current density. It is worth noticing that the total current density $j_T(x,t)$ is given by

$$j_T(x,t) = -\frac{1}{\mu_0 R_0 a^2 x} \frac{\partial}{\partial x}\left(x \frac{\partial \psi}{\partial x}(x,t)\right). \quad (11.3)$$

It allows one to represent the principal term $\frac{\eta_{\|}(x,t)}{\mu_0 a^2} \frac{1}{x} \frac{\partial}{\partial x}\left(x \frac{\partial \psi}{\partial x}(x,t)\right)$ on the right-hand side of the PDE (11.1) in the form $-\eta_{\|}(x,t) R_0 j_T(x,t)$, similar to that of the additive term $\eta_{\|}(x,t) R_0 j_{ni}(x,t)$ on the same. The adequacy of (11.1) to physically investigated phenomena as well as the solution existence and uniqueness are argued in [24].

The *control objective* is to track a desired safety factor profile $q^{target}(x,t)$, which, according to (11.2), does not depend directly on the desired magnetic flux ψ^{target} but does depend on its spatial derivative $\frac{\partial \psi^{target}}{\partial x}$.

For later use, the state transformation

$$\psi_r(x,t) = \psi(x,t) - \psi(t,1) \quad (11.4)$$

11.1 Modeling and Problem Statement

is introduced to deal with homogeneous boundary conditions. Then, taking into account (11.2), one has

$$\psi_r(x,t) = a^2 B_0 \int_x^1 \frac{r}{q(t,r)} dr, \quad \psi_r^{target}(x,t) = a^2 B_0 \int_x^1 \frac{r}{q^{target}(t,r)} dr.$$

The control variables in the infinite-dimensional setting are the plasma loop voltage $V_0(t)$ and the noninductive current density $j_{ni}(x,t)$. $V_0(t)$ can basically directly be set using the inner poloidal magnetic field coils' voltage, whereas $j_{ni}(x,t)$ is manipulated indirectly, using the lower hybrid power P_{lh} and the wave refractive index N_{lh}. The state equation (11.1), rewritten in terms of ψ_r, reduces to

$$\begin{cases} \frac{\partial \psi_r}{\partial t}(x,t) = \frac{\eta_{\|}(x,t)}{\mu_0 a^2} \frac{1}{x} \frac{\partial}{\partial x}\left(x \frac{\partial \psi_r}{\partial x}(x,t)\right) \\ \quad + \eta_{\|}(x,t) R_0 j_{ni}(x,t) + V_0(t), \\ \frac{\partial \psi_r}{\partial x}(x,t)\Big|_{x=0} = 0, \quad \psi_r(t,1) = 0 \end{cases} \quad (11.5)$$

Let's introduce the error variable

$$\phi(x,t) = \psi_r(x,t) - \psi_r^{target}(x) \quad (11.6)$$

with respect to a target $\psi_r^{target}(x)$ that has to be reached in the inner-product space

$$\mathbf{W} = \Big\{ \Psi : (0,1) \to \mathbb{R}, \; \int_0^1 \Psi(x)^2 x \, dx < \infty, \\ \int_0^1 \left|\frac{\partial \Psi(x)}{\partial x}\right|^2 x \, dx < \infty \text{ and } \frac{\partial \Psi}{\partial x}\Big|_{x=0} = \Psi(1) = 0 \Big\} \quad (11.7)$$

of differentiable functions equipped with the inner product $< \Psi_1, \Psi_2 >_\mathbf{W} = \int_0^1 \Psi_1(x) \Psi_2(x) x \, dx$, inducing the norm $\|\Psi\|_\mathbf{W} = \sqrt{\int_0^1 |\Psi(x)|^2 x \, dx}$. As a matter of fact, the target $\psi_r^{target}(x)$, designed a priori, using the manipulatable inputs V_0 and $j_{ni} = j_{ni}(P_{lh}, N_{lh})$, should meet the same homogeneous boundary conditions

$$\frac{\partial \psi_r^{target}}{\partial x}(x)\Big|_{x=0} = 0, \quad \psi_r^{target}(1) = 0 \quad (11.8)$$

as those in (11.5).

The error variable is thus governed by

$$\begin{cases} \frac{\partial \phi}{\partial t}(x,t) = \frac{\eta_{\|}(x,t)}{\mu_0 a^2} \frac{1}{x} \frac{\partial}{\partial x}\left(x \frac{\partial \phi}{\partial x}(x,t)\right) \\ \quad + \frac{\eta_{\|}(x,t)}{\mu_0 a^2} \frac{1}{x} \frac{\partial}{\partial x}\left(x \frac{\partial \psi_r^{target}}{\partial x}(x)\right) \\ \quad + \eta_{\|}(x,t) R_0 j_{ni}(x,t) + V_0(t); \\ \frac{\partial \phi}{\partial x}(x,t)\Big|_{x=0} = 0, \quad \phi(t,1) = 0. \end{cases} \quad (11.9)$$

In order to deal with the regular additive terms on the right-hand side of the PDE (11.9), one should assume that

$$\frac{1}{x}\frac{\partial}{\partial x}\left(x\frac{\partial \psi_r^{target}}{\partial x}(x)\right) \in \mathbf{W}. \qquad (11.10)$$

Let's now introduce the following term

$$\eta_{\|}(x,t) R_0 j_{control} = \eta_{\|}(x,t) R_0 j_{ni}(x,t) + V_0(t) \qquad (11.11)$$

in order to subsequently synthesize a stabilizing control law. The noninductive current density $j_{ni}(x,t)$ is composed of the bootstrap current density $j_{bs}(x,t)$, which is self-generated by the plasma (one can refer to [63] for details) and of additional source terms provided by different actuators, namely, the lower hybrid, electron cyclotron, or ion cyclotron wave systems, and/or the neutral beam injection system. The most efficient is the lower hybrid current drive system that is routinely used on the Tore Supra tokamak [107], which now has a capability to inject up to approximately 7 MW in steady state that will allow plasma currents in the megaampere range to be sustained on a very long pulse [17]. The distributed control law design (proportional and proportional-integral) is generic to all kinds of current drive systems. The numerical simulations will, however, be performed on Tore Supra typical conditions, using mainly lower hybrid waves as the current drive means, so that in this particular case,

$$j_{ni}(x,t) = j_{bs}(x,t) + j_{lh}(x,t), \qquad (11.12)$$

where $j_{bs}(x,t)$ and $j_{lh}(x,t)$ are, respectively, the bootstrap current density and the current density provided by the lower hybrid current drive system. $j_{bs}(x,t)$ can be modeled by a nonlinear function of the flux and pressure [63]. $j_{lh}(x,t)$ is modeled by Gaussian functions controlled by two engineering parameters, the lower hybrid power P_{lh} and the wave refractive index N_{lh} (see [134] for further details). In the subsequent simulations, the variables V_0, P_{lh}, and N_{lh} are viewed as the time-varying control inputs.

Throughout, the resistivity $\eta_{\|}(x,t)$ is assumed to be lower- and upper-bounded by some positive constants η_1 and η_2, namely,

$$\eta_1 \leq \eta_{\|}(x,t) \leq \eta_2 \qquad (11.13)$$

for all $x \in [0,1]$ and $t \geq 0$. Apart from this, $\eta_{\|}(x,t)$ is assumed to be differentiable in t for all $x \in [0,1]$ and its time derivative is assumed to be uniformly bounded:

$$\left|\frac{\partial \eta_{\|}(x,t)}{\partial t}\right| < \Delta \qquad (11.14)$$

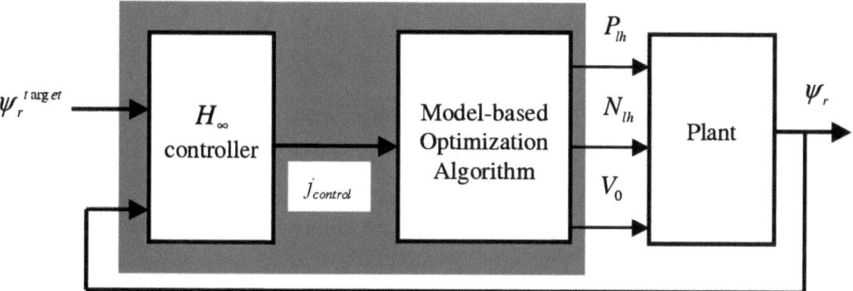

Fig. 11.2 Control scheme of the system

by some constant $\Delta > 0$. Moreover, $\eta_\|(x,t)$ is considered to be available for feedback purposes through some online estimation, basically from electronic temperature measurements (see [134] for more details).

The feedback-control strategy adopted is composed of two steps, as shown in Fig. 11.2. The first step is to synthesize the current density $j_{control}$ to be applied to the system. The resulting $\mathcal{H}_\infty$ controller is designed in the infinite-dimensional setting, where a target flux profile ψ_r^{target} and current flux profile serve as inputs. The second step consists of an optimization process that finds the best set of engineering inputs $P_{lh}(t)$, $N_{lh}(t)$, and $V_0(t)$ to fit, as closely as possible, the desired current density control $j_{control}$ generated by the $\mathcal{H}_\infty$ control law in the infinite-dimensional setting. The optimization algorithm has to minimize the criterion

$$\varepsilon = \int_0^1 \left(j_{control} - j_{engineering}\left(P_{lh}, N_{lh}, V_0 \right) \right)^2 dx \quad (11.15)$$

under the constraints $V_{min} < V_0 < V_{max}$, $P_{min} < P_{lh} < P_{max}$, and $N_{min} < N_{lh} < N_{max}$. $j_{engineering}$ is the control profile that can actually be generated with the real plant control variables P_{lh}, N_{lh}, and V_0 [95]. In simulations presented in Sect. 11.4, the minimum/maximum value of engineering inputs are taken from the Tore Supra tokamak constraints, that is, $-5V < V_0 < 5V$, $0MW < P_{lh} < 7MW$, and $1.43 < N_{lh} < 2.37$.

In the sequel, two distributed controllers are presented.

11.2 Proportional State-Feedback Synthesis

The main goal of this section is to show that the underlying system is exponentially stabilizable using a proportional feedback. The controller specification (11.11) simplifies (11.9) to the PDE

$$\begin{cases} \frac{\partial \phi}{\partial t}(x,t) = \frac{\eta_{\|}(x,t)}{\mu_0 a^2} \frac{1}{x} \frac{\partial}{\partial x}\left(x \frac{\partial \phi}{\partial x}(x,t)\right) \\ \quad + \frac{\eta_{\|}(x,t)}{\mu_0 a^2} \frac{1}{x} \frac{\partial}{\partial x}\left(x \frac{\partial \psi_r^{target}}{\partial x}(x)\right) \\ \quad + \eta_{\|}(x,t) R_0 j_{control}, \\ \frac{\partial \phi}{\partial x}(x,t)\Big|_{x=0} = 0, \quad \phi(t,1) = 0 \end{cases} \quad (11.16)$$

with homogeneous boundary conditions. The control objective is to synthesize $j_{control}$ such that the error state ϕ is exponentially stable.

11.2.1 Disturbance-Free Stabilization

The feedback controller

$$\eta_{\|}(x,t) R_0 j_{control} = -k\phi(x,t) \\ - \frac{\eta_{\|}(x,t)}{\mu_0 a^2} \frac{1}{x} \frac{\partial}{\partial x}\left(x \frac{\partial \psi_r^{target}}{\partial x}(x)\right) \quad (11.17)$$

is proposed to exponentially stabilize the error system (11.16). The closed-loop system (11.16), driven by (11.17), takes the form

$$\begin{cases} \frac{\partial \phi}{\partial t}(x,t) = \frac{\eta_{\|}(x,t)}{\mu_0 a^2} \frac{1}{x} \frac{\partial}{\partial x}\left(x \frac{\partial \phi}{\partial x}(x,t)\right) - k\phi(x,t), \\ \frac{\partial \phi}{\partial x}(x,t)\Big|_{x=0} = 0, \quad \phi(t,1) = 0. \end{cases} \quad (11.18)$$

A technical lemma is involved for later stability analysis.

Lemma 13. *Let $L(\cdot) \in W$ and let $w(\cdot)$ be a Lebesgue-measurable function on $[0,1]$ such that the relations $w_1 \leq w(x) \leq w_2$ are satisfied for all $x \in [0,1]$ and some positive constants w_1, w_2. Then the inequality*

$$\int_0^1 |L(x)|^2 \frac{x}{w(x)} dx \leq K \int_0^1 \left|\frac{\partial L(x)}{\partial x}\right|^2 \frac{x}{w(x)} dx \quad (11.19)$$

holds with the positive constant $K = \frac{w_2}{ew_1}$.

Proof. It is clear that

$$L(1) - L(x) = \int_x^1 \frac{\partial L(s)}{\partial s} ds. \quad (11.20)$$

Since $L \in W$, it follows from (11.7) that $L(1) = 0$ and hence,

$$-L(x) = \int_x^1 \frac{\partial L(s)}{\partial s} ds. \quad (11.21)$$

11.2 Proportional State-Feedback Synthesis

Then one derives

$$\begin{aligned}|L(x)|^2 &= \left|\int_x^1 \sqrt{\tfrac{s}{w(s)}}\tfrac{\partial L(s)}{\partial s}\tfrac{\sqrt{w(s)}}{\sqrt{s}}ds\right|^2\\ &\le \left(\int_x^1 \left|\tfrac{\partial L(s)}{\partial s}\right|^2 \tfrac{s}{w(s)}ds\right)\int_x^1 \tfrac{w(s)}{s}ds\\ &\le \left(\int_x^1 \left|\tfrac{\partial L(s)}{\partial s}\right|^2 \tfrac{s}{w(s)}ds\right)w_2(\ln(1)-\ln x)\\ &\le \left(\int_x^1 \left|\tfrac{\partial L(s)}{\partial s}\right|^2 \tfrac{s}{w(s)}ds\right)w_2(-\ln x). \end{aligned} \tag{11.22}$$

Thus, it leads to

$$\begin{aligned}\tfrac{x}{w(x)}|L(x)|^2 &\le \left(\int_x^1 \left|\tfrac{\partial L(s)}{\partial s}\right|^2 \tfrac{s}{w(s)}ds\right)\tfrac{w_2}{w(x)}(-x\ln x)\\ &\le \left(\int_x^1 \left|\tfrac{\partial L(s)}{\partial s}\right|^2 \tfrac{s}{w(s)}ds\right)\tfrac{w_2}{w_1}(-x\ln x)\\ &\le K\int_x^1 \left|\tfrac{\partial L(s)}{\partial s}\right|^2 \tfrac{s}{w(s)}ds \end{aligned} \tag{11.23}$$

with $K = \tfrac{w_2}{w_1}\sup_{x\in(0,1)}\{-x\ln x\} = \tfrac{w_2}{w_1 e}$. Thus, the required inequality (11.19) is established.

Theorem 41. *Consider the error system (11.16) with the above assumptions, and let (11.16) be driven by controller (11.17). Then the closed-loop system (11.18) is globally exponentially stable with respect to the induced norm* $\|.\|_W$ *provided that the controller gain k is tuned according to*

$$k > -\frac{\eta_1^2 e}{\mu_0 a^2 \eta_2} + \frac{\Delta}{2\eta_1}. \tag{11.24}$$

Proof. Consider the following time-varying functional:

$$V(\phi(t,\cdot),t) = \int_0^1 \phi(x,t)^2 \frac{x}{\eta_\|(x,t)}dx, \tag{11.25}$$

which is positive definite and radially unbounded due to

$$\eta_2^{-1}\|\phi\|_W \le V(\phi(t,\cdot),t) \le \eta_1^{-1}\|\phi\|_W, \tag{11.26}$$

where η_1 and η_2 are the lower and upper bounds in (11.13). The time derivative of (11.25) is given by

$$\begin{aligned}\tfrac{dV(\phi(t,\cdot),t)}{dt} &= \int_0^1 \tfrac{\partial}{\partial t}\left(\phi(x,t)^2 \tfrac{x}{\eta_\|(x,t)}\right)dx\\ &= 2\int_0^1 \tfrac{\partial \phi(x,t)}{\partial t}\phi(x,t)\tfrac{x}{\eta_\|(x,t)}dx\\ &\quad -\int_0^1 \phi(x,t)^2 \tfrac{x}{\eta_\|(x,t)}\tfrac{\partial \eta_\|(x,t)}{\partial t}\tfrac{1}{\eta_\|(x,t)}dx. \end{aligned} \tag{11.27}$$

While being computed along the solutions of (11.18), the time derivative (11.27) integrates by parts to

$$\begin{aligned}\frac{dV(\phi(t,\cdot),t)}{dt} &= 2\int_0^1 \frac{\partial \phi(x,t)}{\partial t}\phi(x,t)\frac{x}{\eta_{\|}(x,t)}dx \\ &\quad - \int_0^1 \phi(x,t)^2 \frac{x}{\eta_{\|}(x,t)} \frac{\frac{\partial \eta_{\|}(x,t)}{\partial t}}{\eta_{\|}(x,t)} dx \\ &= -2k\int_0^1 \phi(x,t)^2 \frac{x}{\eta_{\|}(x,t)} dx \\ &\quad + \frac{2}{\mu_0 a^2}\int_0^1 \frac{\partial}{\partial x}\left(x\frac{\partial \phi(x,t)}{\partial x}\right)\phi(x,t)dx \\ &\quad - \int_0^1 \phi(x,t)^2 \frac{x}{\eta_{\|}(x,t)} \frac{\frac{\partial \eta_{\|}(x,t)}{\partial t}}{\eta_{\|}(x,t)} dx \\ &= -2k\int_0^1 \phi(x,t)^2 \frac{x}{\eta_{\|}(x,t)} dx \\ &\quad - \frac{2}{\mu_0 a^2}\int_0^1 \left|\frac{\partial \phi(x,t)}{\partial x}\right|^2 x\, dx \\ &\quad - \int_0^1 \phi(x,t)^2 \frac{x}{\eta_{\|}(x,t)} \frac{\frac{\partial \eta_{\|}(x,t)}{\partial t}}{\eta_{\|}(x,t)} dx, \end{aligned} \qquad (11.28)$$

thereby yielding

$$\begin{aligned}\frac{dV(\phi(t,\cdot),t)}{dt} &= -2k\int_0^1 \phi(x,t)^2 \frac{x}{\eta_{\|}(x,t)}dx \\ &\quad - \frac{2}{\mu_0 a^2}\int_0^1 \eta_{\|}(x,t)\left|\frac{\partial \phi(x,t)}{\partial x}\right|^2 \frac{x}{\eta_{\|}(x,t)}dx \\ &\quad - \int_0^1 \phi(x,t)^2 \frac{x}{\eta_{\|}(x,t)} \frac{\frac{\partial \eta_{\|}(x,t)}{\partial t}}{\eta_{\|}(x,t)} dx \\ &\leq -2k\int_0^1 \phi(x,t)^2 \frac{x}{\eta_{\|}(x,t)}dx \\ &\quad - \frac{2\eta_1}{\mu_0 a^2}\int_0^1 \left|\frac{\partial \phi(x,t)}{\partial x}\right|^2 \frac{x}{\eta_{\|}(x,t)}dx \\ &\quad - \int_0^1 \phi(x,t)^2 \frac{x}{\eta_{\|}(x,t)} \frac{\frac{\partial \eta_{\|}(x,t)}{\partial t}}{\eta_{\|}(x,t)} dx. \end{aligned} \qquad (11.29)$$

If we specify inequality (11.19) with $w(\cdot) = \eta_{\|}(x,t)$ and $L(\cdot) = \phi(x,t)$, the resulting inequality,

$$\int_0^1 |\phi(x,t)|^2 \frac{x}{\eta_{\|}(x,t)} dx \leq K \int_0^1 \left|\frac{\partial \phi(x,t)}{\partial x}\right|^2 \frac{x}{\eta_{\|}(x,t)} dx, \qquad (11.30)$$

which comes with $K = \frac{\eta_2}{\eta_1 e}$, allows us to conclude from (11.29) that

$$\begin{aligned}\frac{1}{2}\frac{dV(\phi(t,\cdot),t)}{dt} &\leq -\left(\frac{\eta_1^2 e}{\mu_0 a^2 \eta_2} + k\right)\int_0^1 \phi(x,t)^2 \frac{x}{\eta_{\|}(x,t)}dx \\ &\quad + \frac{1}{2}\int_0^1 \phi(x,t)^2 \frac{x}{\eta_{\|}(x,t)} \frac{\left|\frac{\partial \eta_{\|}(x,t)}{\partial t}\right|}{\eta_{\|}(x,t)} dx.\end{aligned} \qquad (11.31)$$

Taking (11.14) into account, we see that

$$\begin{aligned}\frac{1}{2}\frac{dV(\phi(t,\cdot),t)}{dt} &\leq -\left(\frac{\eta_1^2 e}{\mu_0 a^2 \eta_2} + k - \frac{\Delta}{2\eta_1}\right)\int_0^1 \phi(x,t)^2 \frac{x}{\eta_{\|}(x,t)}dx \\ &= -\left(\frac{\eta_1^2 e}{\mu_0 a^2 \eta_2} + k - \frac{\Delta}{2\eta_1}\right)V(\phi(t,\cdot),t). \end{aligned} \qquad (11.32)$$

11.2 Proportional State-Feedback Synthesis

Then

$$V(\phi(t,\cdot),t) \leq V(\phi(0,\cdot),0)e^{-2ct}, \tag{11.33}$$

where $c = \left(\frac{\eta_1^2 e}{\mu_0 a^2 \eta_2} + k - \frac{\Delta}{2\eta_1}\right) > 0$ due to (11.24). By virtue of (11.26), the global exponential stability of the closed-loop system (11.18) is then concluded in the state space **W**.

11.2.2 Disturbance Attenuation

The next goal is to show that the controller defined by (11.17) solves an $\mathcal{H}_\infty$-control problem with some disturbance attenuation level γ. For this purpose, let's consider its perturbed version along with system (11.18):

$$\begin{cases} \frac{\partial \phi}{\partial t}(x,t) = \frac{\eta_{\|}(x,t)}{\mu_0 a^2} \frac{1}{x} \frac{\partial}{\partial x}\left(x \frac{\partial \phi}{\partial x}(x,t)\right) - k\phi(x,t) \\ \qquad\qquad + h(x,t), \\ \frac{\partial \phi}{\partial x}(x,t)\Big|_{x=0} = 0, \quad \phi(t,1) = 0, \end{cases} \tag{11.34}$$

where $h \in L_2(0,\infty;\mathbf{W})$ is an external disturbance, affecting the system. For convenience of the reader, recall that $L_2(0,\infty;\mathbf{W})$ denotes the Hilbert space of square-integrable Hilbert space-valued functions $h(\cdot,x) \in L_2(0,\infty)$ with values $h(t,\cdot) \in \mathbf{W}$ for almost all $t \in [0,\infty)$.

To attenuate external disturbances with a certain level γ, the controller gain k, apart from the global asymptotic stabilization of the nominal system (11.18), has to additionally yield the negative definite performance index (see, e.g., [39] for details):

$$\mathcal{J} = \int_0^\infty [\|\phi(t,\cdot)\|_\mathbf{W}^2 - \gamma^2 \|h(t,\cdot)\|_\mathbf{W}^2] dt \\ - \gamma^2 \|\phi(0,\cdot)\|_\mathbf{W}^2 < 0 \tag{11.35}$$

on the solutions of the perturbed system (11.34) for any $h \in L_2(0,\infty;\mathbf{W})$ and some constant γ.

To solve the problem, let's carry out conditions that guarantee the negative definiteness of the form

$$\mathcal{W}(\phi(t,\cdot),h(t,\cdot)) := \eta_1 \gamma^2 \frac{dV(\phi(t,\cdot))}{dt} \\ + \|\phi(t,\cdot)\|_\mathbf{W}^2 - \gamma^2 \|h(t,\cdot)\|_\mathbf{W}^2 < 0 \tag{11.36}$$

computed on the solutions of the perturbed system (11.34) with $V(\phi(t,\cdot))$ given by (11.25). Once (11.36) is guaranteed, integrating $\mathcal{W}(\phi(t,\cdot),h(t,\cdot))$ in t from 0 to ∞ and taking into account that $\limsup_{t\to\infty} V(\phi(t,\cdot)) \geq 0$ result in (11.35). Indeed, (11.36) ensures that

$$-\eta_1\gamma^2\int_0^1 \phi(0,x)^2\frac{x}{\eta_{\|}(0,x)}dx +$$
$$\int_0^\infty [\|\phi(t,\cdot)\|_W^2 - \gamma^2\|h(t,\cdot)\|_W^2 dt] < 0; \tag{11.37}$$

due to (11.13), the negative definite performance (11.35) follows.

In order to ensure (11.36), let's set $\zeta(x,t) = (|\phi(x,t)|, |h(x,t)|)^T$ and compute the time derivative of (11.25) along the solutions of the perturbed system (11.34). Similar to (11.31), we obtain

$$\begin{aligned}\mathcal{W} &\leq \left[1 - 2\tfrac{\eta_1}{\eta_2}\gamma^2\left(\tfrac{\eta_1^2 e}{\mu_0 a^2 \eta_2} + k - \tfrac{\Delta}{2\eta_1}\right)\right]\int_0^1 \phi(x,t)^2 x dx \\ &+ 2\gamma^2\int_0^1 |h(x,t)||\phi(x,t)|x dx \\ &-\gamma^2\int_0^1 h(x,t)^2 x dx \leq \int_0^1 \zeta^T(x,t)\Phi_\gamma\zeta(x,t) x dx,\end{aligned} \tag{11.38}$$

where

$$\Phi_\gamma := \begin{pmatrix} 1 - 2\tfrac{\eta_1}{\eta_2}\gamma^2\left(\tfrac{\eta_1^2 e}{\mu_0 a^2 \eta_2} + k - \tfrac{\Delta}{2\eta_1}\right) & \gamma^2 \\ \gamma^2 & -\gamma^2 \end{pmatrix}. \tag{11.39}$$

If $\Phi_\gamma < 0$, then $\mathcal{W} < 0$. In turn, due to the Schur complements' formula, the LMI $\Phi_\gamma < 0$ is satisfied if

$$1 - \frac{2\eta_1\gamma^2}{\eta_2}\left(\frac{\eta_1^2 e}{\mu_0 a^2 \eta_2} + k - \frac{\Delta}{2\eta_1}\right) + \gamma^2 < 0.$$

Thus, the tuning rule

$$k > \frac{(1+\gamma^2)\eta_2}{2\gamma^2\eta_1} - \frac{e\eta_1^2}{\mu_0 a^2\eta_2} + \frac{\Delta}{2\eta_1}, \tag{11.40}$$

imposed on the parameter k, ensures the negative definiteness of Φ_γ. The following result can be concluded.

Theorem 42. *Let the conditions of Theorem 41 be satisfied and let the tuning rule (11.40) be additionally imposed on the controller gain k. Then the disturbance-free system (11.34) with $h \equiv 0$ is globally exponentially stable. Moreover, an admissible external disturbance $h \in L_2(0,\infty; W)$, affecting system (11.34), is attenuated in the sense of (11.35).*

Proof. The first assertion is established by Theorem 41. Since relation (11.40) is in force, it follows that (11.39) is negative definite and by means of (11.38), it results in (11.36). Then integrating (11.36) in t from zero to infinity and employing that V is positive definite yield (11.37), thereby concluding (11.35). The second assertion is thus established and the proof of Theorem 42 is completed.

11.3 Proportional-Integral State-Feedback Synthesis

The main goal of this section is to design a controller that would completely reject time-invariant disturbances by using proportional-integral feedback. For this purpose, it is additionally assumed that plasma resistivity is representable in the form

$$\eta_{\|}(x,t) = f(t)g(x) \tag{11.41}$$

subject to the constraints

$$0 < f_1 \leq f(t) \leq f_2, \quad 0 < g_1 \leq g(x) \leq g_2, \tag{11.42}$$

with some constants f_1, f_2, g_1, g_2. Such a realistic assumption is justified in [93, 94, 138].

Next, the inner-product space

$$\mathbf{W}_g = \left\{ \Psi : (0,1) \to \mathbb{R}, \ \int_0^1 \Psi(x)^2 \frac{x}{g(x)} dx < \infty, \right.$$
$$\left. \int_0^1 \left| \frac{\partial \Psi(x)}{\partial x} \right|^2 \frac{x}{g(x)} dx < \infty \ \text{and} \ \frac{\partial \Psi}{\partial x} \Big|_{x=0} = \Psi(1) = 0 \right\} \tag{11.43}$$

of differentiable functions equipped with the inner product

$$< \Psi_1, \Psi_2 >_{\mathbf{W}_g} = \int_0^1 \Psi_1(x) \Psi_2(x) \frac{x}{g(x)} dx$$

and the induced norm $\|\Psi\|_{\mathbf{W}_g} = \sqrt{\int_0^1 |\Psi(x)|^2 \frac{x}{g(x)} dx}$ is defined. It is clear that the space $\mathbf{W}_g$ is nothing other than the previously given space $\mathbf{W}$ equipped with an equivalent norm such that

$$\sqrt{g_2^{-1}} \|\Psi\|_{\mathbf{W}} \leq \|\Psi\|_{\mathbf{W}_g} \leq \sqrt{g_1^{-1}} \|\Psi\|_{\mathbf{W}} \tag{11.44}$$

for any Ψ of either space.

11.3.1 Stabilization Under Time-Invariant Disturbances

Let's now consider the following distributed (proportional-integral feedback) control law:

$$\eta_{\|}(x,t) R_0 j_{control}(x,t) = -k\phi(x,t)$$
$$-k_1 \int_0^t \phi(s,x) ds - \frac{\eta_{\|}(x,t)}{\mu_0 a^2} \frac{1}{x} \frac{\partial}{\partial x} \left(x \frac{\partial \psi_r^{target}}{\partial x}(x) \right), \tag{11.45}$$

where k and k_1 are weight parameters to be defined and j_{control} is extracted from (11.11). Then the error state equation (11.16), driven by the proportional-integral feedback (11.45) and additionally affected by a time-invariant external disturbance $h(x) \in \mathbf{W}_g$, is specified as

$$\begin{cases} \frac{\partial \phi}{\partial t}(x,t) = \frac{\eta_\parallel(x,t)}{\mu_0 a^2} \frac{1}{x} \frac{\partial}{\partial x}\left(x \frac{\partial \phi}{\partial x}(x,t)\right) - k\phi(x,t) \\ \qquad\qquad - k_1 \int_0^t \phi(s,x)ds + h(x), \\ \frac{\partial \phi}{\partial x}(x,t)\Big|_{x=0} = 0, \quad \phi(t,1) = 0. \end{cases} \tag{11.46}$$

In the subsequent analysis, the closed-loop system is shown to be globally asymptotically stable under an arbitrary $h(x) \in \mathbf{W}_g$.

Theorem 43. *Let the controller gains be tuned according to*

$$k > -\frac{f_1 g_1^2 e}{\mu_0 a^2 g_2}, \quad k_1 > 0. \tag{11.47}$$

Then the closed-loop system (11.46) is globally asymptotically stable in the state space $\mathbf{W}_g$ regardless of whichever external disturbance $h(x) \in \mathbf{W}_g$ affects the system; hence, $\phi = 0$ in the steady state.

Proof. For later use, we'll introduce an auxiliary variable $I(x,t)$ governed by

$$\frac{\partial I}{\partial t}(x,t) = \phi(x,t), \quad I(x,0) = 0. \tag{11.48}$$

While being coupled with (11.48), system (11.46), under assumption (11.41), is augmented to

$$\begin{cases} \frac{\partial \phi}{\partial t}(x,t) = \frac{f(t)g(x)}{\mu_0 a^2} \frac{1}{x} \frac{\partial}{\partial x}\left(x \frac{\partial \phi}{\partial x}(x,t)\right) - k\phi(x,t) \\ \qquad\qquad - k_1 I(x,t) + h(x), \\ \frac{\partial I}{\partial t}(x,t) = \phi(x,t), \\ \frac{\partial \phi}{\partial x}(x,t)\Big|_{x=0} = 0, \quad \phi(t,1) = 0. \end{cases} \tag{11.49}$$

The steady state (ϕ_{ss}, I_{ss}) of system (11.49) is clearly defined by $\frac{\partial \phi_{ss}(x,t)}{\partial t} = 0$, $\frac{\partial I_{ss}(x,t)}{\partial t} = 0$, which results in

$$\phi_{ss}(x) = 0, \quad I_{ss}(x) = \frac{h(x)}{k_1}. \tag{11.50}$$

On the solutions of (11.49), consider the functional

$$\begin{aligned} V(\phi(t,\cdot), I(t,\cdot)) &:= \tfrac{1}{2} \int_0^1 \phi(x,t)^2 \tfrac{x}{g(x)} dx \\ &\quad + \tfrac{k_1}{2} \int_0^1 \left(I(x,t) - \tfrac{h(x)}{k_1}\right)^2 \tfrac{x}{g(x)} dx \\ &:= V_1(t) + V_2(t), \end{aligned} \tag{11.51}$$

11.3 Proportional-Integral State-Feedback Synthesis

which proves to be positive definite and radially unbounded. Indeed,

$$V(\phi_{ss}, I_{ss}) = 0, \quad V(\phi, I) > 0 \text{ for any } (\phi, I) \neq (\phi_{ss}, I_{ss}), \quad (11.52)$$

and $V(\phi, I) \to \infty$ as $\|\phi\|_{w_g} + \|I\|_{w_g} \to \infty$. Computing the time derivative

$$\begin{aligned} \frac{dV_1(t)}{dt} &= \frac{1}{2} \int_0^1 \frac{\partial}{\partial t} \left(\phi(x,t)^2 \frac{x}{g(x)} \right) dx \\ &= \int_0^1 \frac{\partial \phi(x,t)}{\partial t} \phi(x,t) \frac{x}{g(x)} dx \end{aligned} \quad (11.53)$$

of (11.51) on the solutions of (11.49) and integrating by parts yield

$$\begin{aligned} \frac{dV_1(t)}{dt} &= \frac{f(t)}{\mu_0 a^2} \int_0^1 \frac{\partial}{\partial x} \left(x \frac{\partial \phi(x,t)}{\partial x} \right) \phi(x,t) dx \\ &\quad -k \int_0^1 \phi(x,t)^2 \frac{x}{g(x)} dx \\ &\quad -k_1 \int_0^1 I(t,x) \phi(x,t) \frac{x}{g(x)} dx \\ &\quad + \int_0^1 h(x) \phi(x,t) \frac{x}{g(x)} dx; \\ &= -\frac{f(t)}{\mu_0 a^2} \int_0^1 \left| \frac{\partial \phi(x,t)}{\partial x} \right|^2 x dx \\ &\quad -k \int_0^1 \phi(x,t)^2 \frac{x}{g(x)} dx - \frac{dV_2(t)}{dt}. \end{aligned} \quad (11.54)$$

It follows that

$$\begin{aligned} \frac{dV_1(t)}{dt} &= -\frac{f(t)}{\mu_0 a^2} \int_0^1 g(x) \left| \frac{\partial \phi(x,t)}{\partial x} \right|^2 \frac{x}{g(x)} dx \\ &\quad -k \int_0^1 \phi(x,t)^2 \frac{x}{g(x)} dx - \frac{dV_2(t)}{dt} \\ &< -\frac{f_1 g_1}{\mu_0 a^2} \int_0^1 \left| \frac{\partial \phi(x,t)}{\partial x} \right|^2 \frac{x}{g(x)} dx \\ &\quad -k \int_0^1 \phi(x,t)^2 \frac{x}{g(x)} dx - \frac{dV_2(t)}{dt}. \end{aligned} \quad (11.55)$$

Taking into account that Lemma 13, which is specified with $w(\cdot) = g(x)$ and $L(\cdot) = \phi(x,t)$, results in

$$\int_0^1 |\phi(x,t)|^2 \frac{x}{g(x)} dx \leq \frac{g_2 e^{-1}}{g_1} \int_0^1 \left| \frac{\partial \phi(x,t)}{\partial x} \right|^2 \frac{x}{g(x)} dx, \quad (11.56)$$

we then arrive at

$$\begin{aligned} \frac{dV(\phi(t,\cdot))}{dt} &= \frac{dV_1(t)}{dt} + \frac{dV_2(t)}{dt} \\ &\leq -\left(\frac{f_1 g_1^2 e}{\mu_0 a^2 g_2} + k \right) \int_0^1 \phi(x,t)^2 \frac{x}{g(x)} dx. \end{aligned} \quad (11.57)$$

Due to (11.47), we can conclude that $\frac{dV(\phi(t,\cdot))}{dt} \leq 0$ and

$$\frac{dV(\phi(t,\cdot))}{dt} = 0 \Rightarrow \int_0^1 \phi(x,t)^2 \frac{x}{g(x)} dx = 0; \quad (11.58)$$

that is, $\phi(x,t) = 0 = \phi_{ss}$ once $\frac{dV(\phi(t,\cdot))}{dt} = 0$. Moreover, relation $\phi(x,t) = 0 = \phi_{ss}$, coupled with (11.48), ensures that $I(x,t) = \frac{h(x)}{k_1} = I_{ss}$, thereby concluding that the maximal invariant manifold of $\frac{dV(\phi(t,\cdot))}{dt} = 0$ coincides with $(\phi, I) = (\phi_{ss}, I_{ss})$. Thus, by applying the LaSalle invariance principle to the parabolic system (11.49) (see [62] for an extension of the invariance principle to parabolic systems), the desired global asymptotic stability is established. Hence, system (11.46) is globally asymptotically stable in $\mathbf{W}$ under an arbitrary $h(x) \in \mathbf{W}_g$.

11.3.2 $\mathcal{H}_\infty$ Design

The objective of this subsection is to show that in addition to the rejection of time-invariant external disturbances, the proposed controller (11.45) attenuates time-varying disturbances with a certain level $\gamma > 0$. Let's suppose that the disturbance h affecting system (11.46) varies in time, thus resulting in the state equation

$$\begin{cases} \frac{\partial \phi}{\partial t}(x,t) = \frac{f(t)g(x)}{\mu_0 a^2} \frac{1}{x} \frac{\partial}{\partial x}\left(x \frac{\partial \phi}{\partial x}(x,t)\right) - k\phi(x,t) \\ \qquad -k_1 \int_0^t \phi(s,x)ds + h(x,t), \\ \frac{\partial \phi}{\partial x}(x,t)\Big|_{x=0} = 0, \quad \phi(t,1) = 0, \end{cases} \tag{11.59}$$

where $h(t,\cdot) \in L_2(0, \infty; \mathbf{W}_g)$ is an external disturbance. It is required to impose an additional condition on the proportional-integral controller gain k that would ensure the negative performance index

$$\begin{aligned} \mathcal{J}_g = \int_0^\infty [\|\phi(t,\cdot)\|_{\mathbf{W}_g}^2 - \gamma^2 \|h(t,\cdot)\|_{\mathbf{W}_g}^2] dt \\ -\gamma^2 \|\phi(0,\cdot)\|_{\mathbf{W}_g}^2 < 0 \end{aligned} \tag{11.60}$$

on the solutions of (11.59) for all time-varying disturbances $h(t,\cdot)$ of class $L_2(0, \infty; \mathbf{W}_g)$ with a constant attenuation level $\gamma > 0$. As before, $L_2(0, \infty; \mathbf{W}_g)$ stands for the Hilbert space of square-integrable Hilbert space-valued functions $h(\cdot, x) \in L_2(0, \infty)$ with values $h(t, \cdot) \in \mathbf{W}_g$ for almost all $t \in [0, \infty)$.

To establish (11.60), it suffices to demonstrate that

$$\begin{aligned} \mathcal{W}_g(\phi(t,\cdot)) := 2\gamma^2 \frac{d}{dt} V_g(\phi(t,\cdot)) \\ + \int_0^1 (\phi(x,t)^2 - \gamma^2 h(x,t)^2) \frac{x}{g(x)} dx < 0, \end{aligned} \tag{11.61}$$

where

$$V_g(\phi(t,\cdot)) = \frac{1}{2} \int_0^1 \phi(x,t)^2 \frac{x}{g(x)} dx + \frac{k_1}{2} \int_0^1 I(x,t)^2 \frac{x}{g(x)} dx. \tag{11.62}$$

Indeed, integrating $\mathcal{W}_g(\phi(t,\cdot))$ in t from 0 to ∞ and taking into account that $\limsup_{t\to\infty} V_g(\phi(\infty,\cdot)) \geq 0$ result in (11.60).

11.3 Proportional-Integral State-Feedback Synthesis

Let's now compute the time derivative of functional (11.62) on the solutions of (11.59):

$$\frac{dV_g(\phi(t,\cdot))}{dt} \leq -\left(\frac{f_1 g_1^2 e}{\mu_0 a^2 g_2} + k\right) \int_0^1 \phi(x,t)^2 \frac{x}{g(x)} dx + \int_0^1 h(x,t)\phi(x,t)\frac{x}{g(x)} dx. \quad (11.63)$$

Setting $\zeta = (|\phi(x,t)| \ |h(x,t)|)^T$, we see that

$$\begin{aligned} \mathcal{W}_g \leq &\left(1 - 2\gamma^2\left(\frac{f_1 g_1^2 e}{\mu_0 a^2 g_2} + k\right)\right)\int_0^1 \phi(x,t)^2 \frac{x}{g(x)} dx \\ &+ 2\gamma^2 \int_0^1 |h(x,t)||\phi(x,t)|\frac{x}{g(x)} dx \\ &- \gamma^2 \int_0^1 |h(x,t)|^2 \frac{x}{g(x)} dx \leq \int_0^1 \zeta^T \Psi_\gamma \zeta \frac{x}{g(x)} dx, \end{aligned} \quad (11.64)$$

where

$$\Psi_\gamma := \begin{pmatrix} 1 - 2\gamma^2\left(\frac{f_1 g_1^2 e}{\mu_0 a^2 g_2} + k\right) & \gamma^2 \\ \gamma^2 & -\gamma^2 \end{pmatrix}. \quad (11.65)$$

If $\Psi_\gamma < 0$, then $\mathcal{W}_g < 0$. By the Schur complements' formula, the LMI condition $\Psi_\gamma < 0$ is satisfied if

$$k > -\frac{f_1 g_1^2 e}{\mu_0 a^2 g_2} + \frac{1}{2\gamma^2} + \gamma^2. \quad (11.66)$$

Hence, $\mathcal{W}_g < 0$ is guaranteed provided that k is chosen according to (11.47) and (11.66). Thus, the following result is in order.

Theorem 44. *Let the conditions of Theorem 43 be satisfied and let the tuning rule (11.66) be additionally imposed on the gain k of the proportional-integral feedback (11.45). Then, while being affected by a time-invariant external disturbance $h(x) \in \mathbf{W}_g$, the closed-loop system (11.45), (11.59) is globally asymptotically stable in the state space $\mathbf{W}_g$. Moreover, all time-varying external disturbances $h(t,\cdot) \in L_2(0,\infty;\mathbf{W}_g)$ are attenuated in the sense of (11.60).*

Proof. The first assertion is established by Theorem 43. By virtue of (11.66), matrix (11.65) is negative definite. Coupled with (11.64), this results in (11.61). If we want to justify (11.60), it suffices to integrate (11.61) in t from zero to infinity and then, in the resulting inequality,

$$\begin{aligned} &\int_0^\infty [\|\phi(t,\cdot)\|_{\mathbf{W}_g}^2 - \gamma^2 \|h(t,\cdot)\|_{\mathbf{W}_g}^2] dt \\ &+ 2\gamma^2 V_g(\phi(\infty)) - 2\gamma^2 V_g(\phi(0)) < 0, \end{aligned} \quad (11.67)$$

to omit the positive definite term $2\gamma^2 V_g(\phi(\infty,\cdot))$ and to take into account that $2\gamma^2 V_g(\phi(0,\cdot)) = \gamma^2 \|\phi(0,\cdot)\|^2_{\mathbf{W}_g}$ by virtue of (11.62) to be initialized in accordance with (11.48). Theorem 44 is thus proved.

11.4 Simulation Results

An illustrative test is performed to regulate the evolution of the magnetic flux ψ_r from an initial profile to a target profile (see Fig. 11.3 for the profiles). First, simulation runs are made to tune the proportional (k) and integral (k_1) gains. In these simulations, the engineering parameters, reconstructed in the block "Model-based Optimization Algorithm" depicted in Fig. 11.2, are not utilized. Instead, the $\mathcal{H}_\infty$ controller $j_{control}$ is straightforwardly applied to the plant. The effect of the disturbance

$$h(t,x) = \begin{cases} 0, & 0s \le t < 10s, \\ -\frac{1}{5\mu_0 R_0 a^2} \frac{1}{x} \frac{\partial}{\partial x}\left(x \frac{\partial \psi_r^{target}}{\partial x}\right), & 10s \le t \le 30s, \end{cases}$$

added at the time instant $t = 10\,\mathrm{s}$ and representing 20 % of the target's total current density $j_T^{target}(t,x)$, given by (11.3), is additionally tested. The behavior of the closed-loop system is illustrated in Fig. 11.4 by the time evolution of ϕ at point $x = 0.4$ (a similar behavior was also obtained at other points x). As theoretically expected, (1) the closed-loop system is asymptotically stable and the applied disturbance is attenuated once the controller (proportional and integral) gains are properly tuned; (2) if confined to the proportional feedback only, the larger the proportional gain that is applied, the smaller the steady-state error, resulting from a disturbance, that is obtained; (3) the time response is sped up by using a higher proportional gain but at the expense of possibly a higher overshoot; (4) the

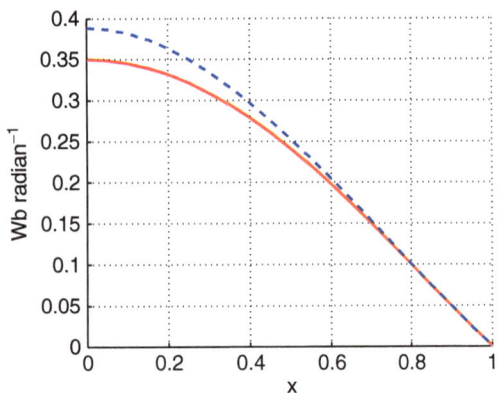

Fig. 11.3 Initial profile value $\psi_r(0,x)$ (*dashed line*) versus the target profile $\psi_r^{target}(x)$ (*solid line*)

11.4 Simulation Results

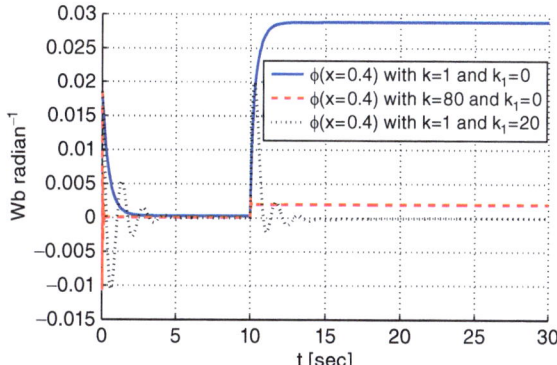

Fig. 11.4 Time evolution of ϕ profile at $x = 0.4$: *Solid line* is for the gain values $(k = 1, k_1 = 0)$; *dashed line* is for $(k = 80, k_1 = 0)$; and *dotted line* is for $(k = 1, k_1 = 20)$

integral gain allows one to cancel the steady-state error resulting from the applied disturbance; higher values of the integral gain speed up the attenuation of the applied disturbance at the expense of induced steady-state oscillations.

In order to further validate the proposed control approach, the tokamak plant simulator METIS (Minute Embedded Tokamak Integrated Simulator) [9] is used. A METIS code includes a full fast current diffusion solver and takes into account various nonlinear couplings between physical quantities. METIS is used jointly with the $MATLAB/Simulink^{TM}$ toolbox to simulate Tore Supra plasma discharges. A preliminary METIS open-loop simulation is performed, using real pulse engineering inputs ($V_0 = 0\,V$, $P_{lh} = 4.8\,MW$ and $N_{lh} = 1.7$), in order to carry out a reachable target profile. The proportional-integral feedback is then implemented using values of k and k_1 resulting from the previously mentioned tuning process, where reasonably low values $k = 1\,\text{s}^{-1}$ and $k_1 = 0.8\,\text{s}^{-2}$ are chosen to avoid deal with high overshooting and induced steady-state oscillations. At $t = 10.1\,\text{s}$, the controller is activated in order to force the magnetic flux profile to reach the target profile. The target profile is basically reached in 5.8 s. As in the preliminary infinite-dimensional simulations, made without engineering parameter optimization, the disturbance $\tilde{h}(t, x) = h(t - 6, x)$, specified with (11.68), is added at $t = 16\,\text{s}$. Simulation results are presented in Figs. 11.5–11.8.

The time evolution of the magnetic flux $\psi_r(t, x)$ versus the target profile $\psi_r^{target}(x)$ at the points $x = 0.2$, $x = 0.4$ and $x = 0.6$, and that of the their integral discrepancy are shown in Figs. 11.5 and 11.6, respectively. Proportional-integral feedback appears to act satisfactorily. Actually, a gentle dynamic behavior is observed and the stationary disturbance is properly attenuated on a sensible time scale. The time evolutions of the engineering plant inputs is plotted in Fig. 11.7. The time evolutions of the relevant control deviation $\varepsilon_{rel}(t, x) = \frac{j_{control}(t,x) - j_{engineering}(t,x)}{j_{control}(t,x)}$ at the points $x = 0.1$ and $x = 0.5$ are shown in Fig. 11.8.

We thus conclude from Figs. 11.5–11.8 that due to the $\mathcal{H}_\infty$ approach used, the essential discrepancy between $j_{engineering}(t, x)$ and $j_{control}(t, x)$ is properly attenuated along with the applied disturbance $\tilde{h}$, so that the target profile is approached in an appropriate time less than 6 s.

Fig. 11.5 Time evolution of ψ_r (*solid line*) versus the target ψ_r^{target} (*dashed line*) at $x = 0.2$ (the *upper plot*), at $x = 0.4$ (the *middle plot*), and at $x = 0.6$ (the *lower plot*)

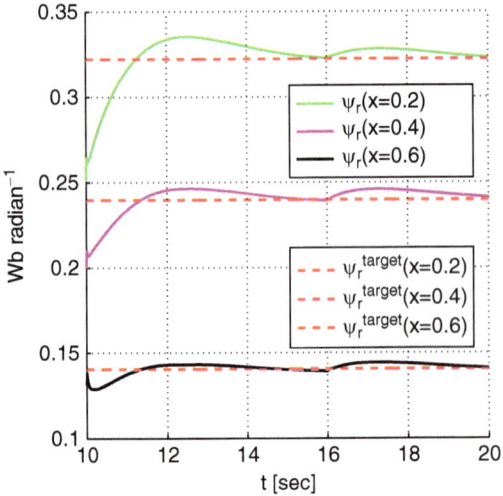

Fig. 11.6 Time evolution of $\int_0^1 \left[\psi_r(t,x) - \psi_r^{target}(x)\right]^2 dx$

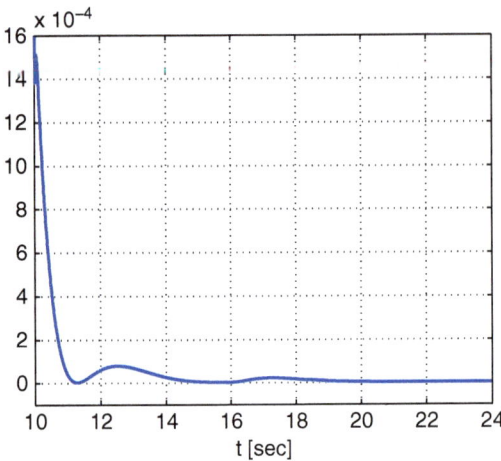

11.4 Simulation Results

Fig. 11.7 Time evolution of the engineering inputs versus their nominal values, used for producing the target profile $\psi_r^{target}(x)$: The *upper plot* is for P_{LH} in megawatts; the *middle plot* is for N_{LH}; the *lower plot* is for V_0 in volts

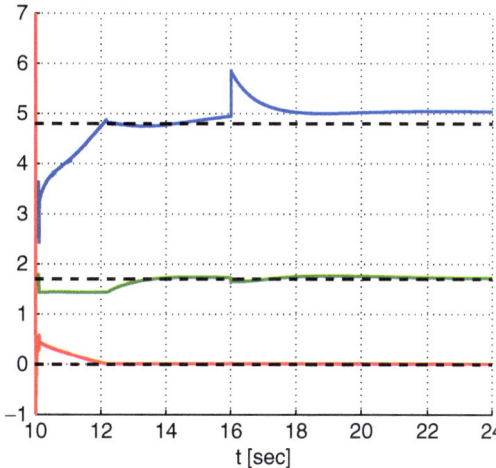

Fig. 11.8 Time evolution of the relevant deviation $\varepsilon_{rel}(t, x) = \frac{j_{control}(t,x) - j_{engineering}(t,x)}{j_{control}(t,x)}$, computed at $x = 0.1$ (*solid line*), and that computed at $x = 0.5$ (*dashed line*)

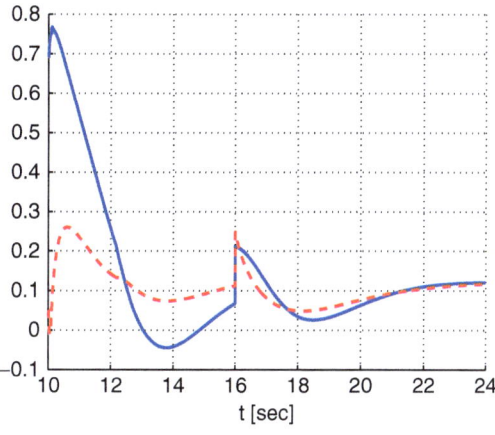

References

1. Acho, L., Orlov, Y., Solis, V.: Nonlinear measurement feedback $\mathcal{H}_\infty$ control of time-periodic systems with application to tracking control of robot manipulator. Int. J. Control **74**(2), 190–198 (2001)
2. Aguilar, L., Orlov, Y., Acho, L.: Nonlinear $\mathcal{H}_\infty$-control of nonsmooth time varying systems with application to friction mechanical manipulators. Automatica **39**, 1531–1542 (2003)
3. Aguilar, L., Orlov, Y., Merzouki, R., Cadiou, J.: Nonlinear $\mathcal{H}_\infty$-output regulation of a multistable drive system with backlash using its single-stability approximation. In: American Control Conference ACC'07, New York, pp. 4709–4714 (2007)
4. Aguilar, L., Orlov, Y., Merzouki, R., Cadiou, J.: Nonlinear $\mathcal{H}_\infty$ output regulation of a nonminimum phase servomechanism with backlash. ASME J. Dyn. Syst. Meas. Control **129**(4), 544–549 (2007)
5. Anderson, B., Vreugdenhil, R.: Network Analysis and Synthesis. Prentice Hall, Englewood Cliffs (1973)
6. Andronov, A., Vitt, A.: On Lyapunov stability. Exp. Theor. Phys. **3**, 373–374 (1933)
7. Arimoto, S.: Control Theory of Non-linear Mechanical Systems: A Passivity-Based and Circuit-Theoretic Approach. Oxford University Press, New York (1996)
8. Armstrong-Hélouvry, B., Dupont, P., Canudas-de-Wit, C.: A survey of models, analysis tools and compensation methods for the control of machines with friction. Automatica **30**, 1083–1138 (1994)
9. Artaud, J.: METIS User's Guide. CEA, IRFM (2008). Technical report
10. Åström, K., Canudas-de-Wit, C., Gafvert, M., Lischinsky, P.: Friction models and friction compensation. Eur. J. Control **4**, 176–195 (1998)
11. Avila-Vilchis, J., Brogliato, B., Dzul, A., Lozano, R.: Nonlinear modelling and control of helicopters. Automatica **39**, 1526–1530 (2003)
12. Balas, G., Chiang, R., Packard, A., Safonov, M.: Robust Control Toolbox. Version 3. Math Works, Natick, MA (2005)
13. Ball, J., Helton, J., Walker, M.: $\mathcal{H}_\infty$-control for nonlinear systems with output feedback. IEEE Trans. Automat. Contr. **38**, 546–559 (1993)
14. Basar, T., Bernhard, P.: $\mathcal{H}_\infty$-Optimal Control and Related Minimax Design Problems: A Dynamic Game Approach, 2nd edn. Birkhauser, Boston (1995)
15. Basar, T., Olsder, G.: Dynamic Noncooperative Game Theory. Academic, New York (1982)
16. Becker, N., Grimm, W.: Nonlinear $\mathcal{H}_\infty$ control of time varying systems: a unified distribution-based formalism for continuous and sample-data measurement feedback design. IEEE Trans. Automat. Contr. **46**(4), 648–643 (2003)
17. Becoulet, A., et al.: Steady state long pulse tokamak operation using lower hybrid current drive. Fusion Eng. Des. **86**(6–8), 490–496 (2011)

18. Bensoussan, A., Prato, G.D., Delfour, M., Mitter, S.: Representation and Control of Infinite Dimensional Systems, vol. 2. Birkhauser, Boston (1993)
19. Bentsman, J., Orlov, Y., Aguilar, L.: Nonsmooth $\mathcal{H}_\infty$ output regulation with application to a coal-fired boiler/turbine unit with actuator deadzone. In: American Control Conference ACC'13, Washington, pp. 3900–3905 (2013)
20. Bentsman, J., Zheng, K., Taft, C.: Advanced boiler/turbine control and its benchmarking in a coal-fired power plant. In: 14th Annual Joint ISA POWID/EPRI Controls and Instrumentation Conference. ISA, Research Triangle Park (2004)
21. Blum, J.: Numerical Simulation and Optimal Control in Plasma Physics with Applications to Tokamaks. Gauthier-Villars Series in Modern Applied Mathematics. Wiley, New York (1989)
22. Boskovic, D.M., Krstic, M.: Stabilization of a solid propellant rocket instability by state feedback. Int. J. Robust Nonlinear Control **13**, 483–495 (2003)
23. Boyd, S., Ghaoui, L.E., Feron, E., Balakrishnan, V.: Linear Matrix Inequality in Systems and Control Theory. Studies in Applied Mathematics. SIAM, Philadelphia (1994)
24. Bribiesca-Argomedo, F., Prieur, C., Witrant, E., Bremond, S.: A strict control Lyapunov function for diffusion equation with time-varying distributed coefficients. IEEE Trans. Automat. Contr. (2012). doi:10.1109/TAC2012.2209260
25. Brogliato, B., Lozano, R., Maschke, B., Egeland, O.: Dissipative Systems Analysis and Control—Theory and Applications, 2nd edn. Springer, London (2007)
26. Canudas-de-Wit, C.: On the concept of virtual constraints as a tool for walking robot control and balancing. Annu. Rev. Control **28**, 157–166 (2004)
27. Canudas-de-Wit, C., Astrom, K., Sorine, M., Olsson, H.: Control of systems with dynamic friction. In: Workshop on Systems with Friction (1999)
28. Chaturvedi, N., McClamroch, N., Bernstein, D.: Stabilization of a 3D axially symmetric pendulum. Automatica **44**(9), 2258–2265 (2008)
29. Chen, B.: Robust and $\mathcal{H}_\infty$ Control. Springer, London (2000)
30. Chung, C., Hauser, J.: Nonlinear $\mathcal{H}_\infty$-control around periodic orbits. Syst. Control Lett. **30**, 127–137 (1997)
31. Clarke, F.: Optimization and Non-smooth Analysis. Wiley Interscience, New York (1983)
32. Colaneri, P.: Continuous-time periodic systems in $\mathcal{H}_2$ and $\mathcal{H}_\infty$—part 1: theoretical aspects. Kybernetika **36**(2), 211–242 (2000)
33. Colaneri, P.: Continuous-time periodic systems in $\mathcal{H}_2$ and $\mathcal{H}_\infty$—part 2: state feedback problems. Kybernetika **36**(3), 329–350 (2000)
34. Crandall, M., Lions, P.: Viscosity solutions of Hamilton–Jacobi equations. Trans. Am. Math. Soc. **277**, 1–42 (1983)
35. Curtain, R., Zwart, H.: An Introduction to Infinite-Dimensional Linear Systems. Springer, New York (1995)
36. Dahl, P.: Solid friction damping of mechanical vibrations. AIAA **14**, 1675–1682 (1976)
37. Datko, R.: Not all feedback stabilized hyperbolic systems are robust with respect to small time delays in their feedbacks. SIAM J. Control Optim. **26**, 697–713 (1988)
38. Doyle, J., Glover, K., Khargonekar, P., Francis, B.: State-space solutions to standard $\mathcal{H}_2$ and $\mathcal{H}_\infty$ control problems. IEEE Trans. Automat. Contr. **34**(8), 831–847 (1989)
39. Du, C., Xie, L.: H_∞ Control and Filtering of Two-Dimensional Systems. Springer, Berlin (2002)
40. Dullerud, G., Paganini, F.: A Course in Robust Control Theory—A Convex Approach. Springer, New York (2002)
41. Dzul, A., Lozano, R., Castillo, P.: Adaptive control for a radio-controlled helicopter in a vertical flying stand. Int. J. Adapt. Control Signal Process. **18**, 473–485 (2004)
42. Filippov, A.: Differential Equations with Discontinuous Right-Hand Sides. Kluwer, Dordrecht (1988)
43. Foias, C., Ozbay, H., Tannenbaum, A.: Robust Control of Infinite Dimensional Systems: Frequency Domain Methods. Springer, London (1996)
44. Francis, B.: A Course in $\mathcal{H}_\infty$ Control Theory. Springer-Verlag, Berlin (1987)

45. Francis, B., Doyle, J.: Linear control theory with an $\mathcal{H}_\infty$ optimality criterion. SIAM J. Control Optim. **25**, 815–844 (1987)
46. Freidovich, L., Mettin, U., Shiriaev, A., Spong, M.: A passive 2-DOF walker: hunting for gaits using virtual holonomic constraints. IEEE Trans. Robot. **25**(5), 1202–1208 (2009)
47. Freidovich, L., Robersson, A., Shiriaev, A., Johansson, R.: Periodic motions of the pendubot via virtual holonomic constraints: theory and experiments. Automatica **44**, 785–791 (2008)
48. Fridman, E.: New Lyapunov–Krasovskii functionals for stability of linear retarded and neutral type systems. Syst. Control Lett. **43**, 309–319 (2001)
49. Fridman, E.: State feedback $\mathcal{H}_\infty$ control of nonlinear singularly perturbed systems. Int. J. Robust Nonlinear Control **11**(12), 1115–1125 (2001)
50. Fridman, E., Orlov, Y.: Exponential stability of linear distributed parameter systems with time-varying delays. Automatica **45**(1), 194–201 (2009)
51. Fridman, E., Orlov, Y.: An LMI approach to $\mathcal{H}_\infty$ boundary control of semilinear parabolic and hyperbolic systems. Automatica **45**(9), 2060–2066 (2009)
52. Fridman, E., Shaked, U.: Delay-dependent stability and $\mathcal{H}_\infty$ control: constant and time-varying delays. Int. J. Control **76**(1), 48–60 (2001)
53. Gabinet, P., Apkarian, P.: A linear matrix inequality approach to $\mathcal{H}_\infty$ control. Int. J. Robust Nonlinear Control **4**(4), 421–448 (1994)
54. Gaye, O., Autrique, L., Orlov, Y., Moulay, E., Bremond, S., Nouailletas, R.: $\mathcal{H}_\infty$ stabilization of the current profile in tokamak plasmas via lmi approach. Automatica (2013). http//dx.doi.org.10.1016/j.automatica.2013.05.011
55. Gouaisbaut, F., Peaucelle, D.: Delay-dependent stability of time delay systems. In: Proceedings of the 5th IFAC Symposium on Robust Control Design, Toulouse (2006)
56. Green, M., Glover, K., Limebeer, D., Doyle, J.: A j-spectral factorization approach to $\mathcal{H}_\infty$ control. SIAM J. Control Optim. **28**, 1350–1371 (1990)
57. Green, M., Limebeer, D.: Linear Robust Control. Prentice Hall, Englewood Cliffs (1995)
58. Gu, K., Kharitonov, V., Chen, J.: Stability of Time-Delay Systems. Birkhauser, Boston (2003)
59. Hale, J., Verduyn Lunel, S.: An Introduction to Functional Differential Equations. Springer, New York (1993)
60. He, Y., Wang, Q.G., Lin, C., Wu, M.: Delay-range-dependent stability for systems with time-varying delay. Automatica **43**(2), 371–376 (2007)
61. Helton, J., James, M.: Extending $\mathcal{H}_\infty$ Control to Nonlinear Systems—Control of Nonlinear Systems to Achieve Performance Objectives. SIAM, Philadelphia (1999)
62. Henry, D.: Geometric Theory of Semilinear Parabolic Equations. Lecture Notes in Mathematics. Springer, New York (1981)
63. Hirshman, S.: Finite-aspect-ratio effects on the bootstrap current in tokamaks. Phys. Fluids **31**(10), 3150–3152 (1998)
64. Huang, C., Vanderwalle, S.: An analysis of delay-dependent stability for ordinary and partial differential equations with fixed and distributed delays. SIAM J. Sci. Comput. **25**(5), 1602–1632 (2004)
65. Ikeda, K., Azuma, T., Uchida, K.: Infinite-dimensional LMI approach to analysis and synthesis for linear time-delay systems. Kibenetika **37**(4), 505–520 (2001)
66. Isidori, A.: Nonlinear Control Systems II. Springer, London (1999)
67. Isidori, A., Astolfi, A.: Disturbance attenuation and $\mathcal{H}_\infty$-control via measurement feedback in nonlinear systems. IEEE Trans. Automat. Contr. **37**(9), 1283–1293 (1992)
68. Isidori, A., Byrnes, C.I.: Output regulation of nonlinear systems. IEEE Trans. Automat. Contr. **35**, 131–140 (1990)
69. Isidori, A., Marconi, L., Serrani, A.: Robust nonlinear motion control of a helicopter. IEEE Trans. Automat. Contr. **48**(3), 413–426 (2003)
70. Kelly, R., Llamas, J., Campa, R.: A measurement procedure for viscous and Coulomb friction. IEEE Trans. Instrum. Meas. **49**, 857–861 (2000)
71. Khalil, H.: Nonlinear Systems, 3rd edn. Prentice Hall, Englewood Cliffs (2002)
72. Klafter, R., Chmielewski, T., Negin, M.: Robotic Engineering: An Integrated Approach. Prentice Hall, Englewood Cliffs (1989)

73. Kolmanovskii, V., Myshkis, A.: Applied Theory of Functional Differential Equations. Kluwer, Dordrecht (1999)
74. Krasnoselskii, M., Zabreyko, P., Pustylnik, E., Sobolevski, P.: Integral Operators in Spaces of Summable Functions. Noordhoff, Leyden (1976)
75. Kwakernaak, H.: A polynomial approach to minimax frequency domain optimization of multivariable feedback systems. Int. J. Control **41**, 117–156 (1986)
76. Lagerberg, A., Egardt, B.: Estimation of backlash with application to automotive powertrains. In: Proceedings of the 42nd Conference on Decision and Control, Maui, pp. 4521–4526 (2003)
77. Leonov, G.: Generalization of Andronov–Vitt theorem. Regul. Chaotic Dyn. **11**(2), 282–289 (2006)
78. Liusternik, L., Sobolev, V.: Elements of Functional Analysis. Frederick Ungar Publishing Company, New York (1961)
79. Logemann, H., Rebarber, R., Weiss, G.: Conditions for robustness and nonrobustness of the stability of feedback systems with respect to small delays in the feedback loop. SIAM J. Control Optim. **34**(2), 572–600 (1996)
80. McEneaney, W.M.: Max-Plus Methods for Nonlinear Control and Estimation. Birkhauser, Boston (2008)
81. Merzouki, R., Cadiou, J.: Compensation of friction and backlash effects in an electrical actuator. J. Syst. Control Eng. **218**, 75–84 (2004)
82. Meza, I., Aguilar, L., Shiriaev, A., Freidovich, L., Orlov, Y.: Periodic motion planning and nonlinear $\mathcal{H}_\infty$ tracking control of a 3-DOF underactuated helicopter. Int. J. Syst. Sci. **42**(5), 829–838 (2011)
83. Mondie, S., Kharitonov, V.: Output regulation of nonlinear systems. IEEE Trans. Automat. Contr. **50**(5), 263–273 (2005)
84. Nicaise, S., Pignotti, C.: Stability and instability results of the wave equation with a delay term in the boundary or internal feedbacks. SIAM J. Control Optim. **45**(5), 1561–1585 (2006)
85. Nicaise, S., Pignotti, C.: Stabilization of the wave equations with variable coefficients and boundary condition of memory type. Asymptot. Anal. **50**(1–2), 31–67 (2006)
86. Niculescu, S.: Delay Effects on Stability: A Robust Control Approach. Lecture Notes in Control and Information Sciences, vol. 269. Springer, London (2001)
87. Nordin, M., Bodin, P., Gutman, P.: New models and identification methods for backlash and gear play. In: Tao, G., Lewis, F.L. (eds.) Adaptive Control of Nonsmooth Dynamic Systems, pp. 1–30. Springer, London (2001)
88. Orlov, Y.: Regularization of singular $\mathcal{H}_2$ and $\mathcal{H}_2$ control problems. In: Proceedings of 36th Conference on Decision and Control, San Diego, pp. 4140–4144 (1997)
89. Orlov, Y.: Discontinuous Systems: Lyapunov Analysis and Robust Synthesis Under Uncertainity Conditions. Springer, London (2009)
90. Orlov, Y., Acho, L., Solis, V.: Nonlinear $\mathcal{H}_\infty$-control of time varying systems. In: Proceedings of 38th Conference on Decision and Control, Phoenix, pp. 3764–3769 (1999)
91. Orlov, Y., Aguilar, L.: Nonsmooth $\mathcal{H}_\infty$ position control of mechanical manipulators with frictional joints. Int. J. Control **77**(11), 1062–1069 (2004)
92. Ortega, R., Loria, A., Nicklasson, P., Sira-Ramírez, H.: Passivity-Based Control of Euler–Lagrange Systems. Springer, London (1998)
93. Ou, Y., Xu, C., Schuster, E., Ferron, J., Luce, T., Walker, M., Humphreys, D.: Receding-horizon optimal control of the current profile evolution during the ramp-up phase of a tokamak discharge. Control Eng. Pract. **19**(1), 22–31 (2011)
94. Ou, Y., Xu, C., Schuster, E., Luce, T., Ferron, J., Walker, M., Humphreys, D.: Optimal tracking control of current profile in tokamaks. IEEE Trans. Control Syst. Technol. **19**(2), 432–441 (2011)
95. Ouarit, H., Bremond, S., Nouailletas, R., Witrant, E., Autrique, L.: Validation of plasma current profile model predictive control in tokamaks via simulations. Fusion Eng. Des. **86**, 1018–1021 (2011)

96. Pazy, A.: Semigroups of Linear Operators and Applications to Partial Differential Equations. Springer, New York (1983)
97. Petersen, I.: Disturbance attenuation and $\mathcal{H}_\infty$ optimization: a design method based on the algebraic Riccati equation. IEEE Trans. Automat. Contr. **32**, 427–429 (1987)
98. Petersen, I., Ugrinovskii, V., Savkin, A.: Robust Control Design Using $\mathcal{H}_\infty$ Methods. Springer, London (2000)
99. Quanser: 3D helicopter system with active disturbance. Tech. rep. (2004)
100. Ravi, R., Nagpal, K., Khargonekar, P.: $\mathcal{H}_\infty$ control of linear time-varying systems: a state-space approach. SIAM J. Control Optim. **29**, 1394–1413 (1991)
101. Rebarber, R., Townley, S.: Robustness with respect to delays for exponential stability of distributed parameter systems. SIAM J. Control Optim. **37**(1), 230–244 (1998)
102. Rebiai, S., Zinober, A.: Stabilization of uncertain distributed parameter systems. Int. J. Control **57**(5), 1167–1175 (1993)
103. Richard, J.P.: Time-delay systems: an overview of some recent advances and open problems. Automatica **39**, 1667–1694 (2003)
104. Rouche, N., Habets, P., Laloy, M.: Stability Theory by Liapunov's Direct Method. Applied Mathematical Sciences, vol. 22. Springer, New York (1977)
105. Safonov, M., Jonckhere, E., Verma, M., Limebeer, D.: Synthesis of positive real multivariable feedback systems. Int. J. Control **45**(3), 817–842 (1987)
106. Safonov, M., Limebeer, D., Chang, R.: Simplifying the $\mathcal{H}_\infty$ theory via loop shifting, matrix pencil, and descriptor concepts. Int. J. Control **50**, 2467–2488 (1989)
107. Saoutic, B., Chatelier, M., Michelis, C.D.: Tore Supra: toward steady state in a superconducting tokamak. Fusion Sci. Technol. **56**(3), 1079–1091 (2009)
108. Shahian, B., Hassul, M.: Control System Design Using MATLAB. Prentice Hall, Englewood Cliffs (1993)
109. Shayman, M.: Inertia theorems for the periodic Liapunov equation and periodic Riccati equations. Syst. Control Lett. **4**, 27–32 (1984)
110. Shiriaev, A., Freidovich, L., Gusev, S.: Transverse linearization for controlled mechanical systems with severe passive degrees of freedom. IEEE Trans. Automat. Contr. **55**(4), 893–906 (2010)
111. Shiriaev, A., Freidovich, L., Manchester, I.: Can we make a robot ballerina perform a pirouette? Orbital stabilization of periodic motions of underactuated mechanical systems. Annu. Rev. Control **32**(2), 200–211 (2008)
112. Shiriaev, A., Perram, J., Canudas-de-Wit, C.: Contructive tool for an orbital stabilization of underactuated nonlinear systems: virtual constraint approach. IEEE Trans. Automat. Contr. **50**(8), 1164–1176 (2005)
113. Shiriaev, A., Robertsson, A., Perram, J., Sandberg, A.: Periodic motion planning for virtually constrained Euler Lagrange systems. Syst. Control Lett. **55**(11), 900–907 (2006)
114. Skogestad, S., Postlethwaite, I.: Multivariable Feedback Control—Analysis and Design. Wiley, Chichester (2005)
115. Spong, M., Hutchinson, S., Vidyasagar, M.: Robot Modeling and Control. Wiley, New York (2006)
116. Stoorvogel, A., Trextelman, H.: The quadratic matrix inequality insingular $\mathcal{H}_\infty$ control with state feedback. SIAM J. Control Optim. **28**, 1190–1208 (1990)
117. Subbotin, A.: Generalized Solutions of First-Order PDE's. Birkhauser, Boston (1995)
118. Sun, X.M., Zhao, J., Hill, D.: Stability and l_2-gain analysis for switched delay systems. Automatica **42**, 1769–1774 (2003)
119. Tadmor, G.: Worst-case design in the time domain: the maximum principle and the standard $\mathcal{H}_\infty$ problem. Math. Control Signals Syst. **3**, 301–324 (1990)
120. Tadmor, G.: The standard $\mathcal{H}_\infty$ problem and the maximum principle: the general linear case. SIAM J. Control Optim. **31**(4), 813–846 (1993)
121. Van der Shaft, A.: A state-space approach to nonlinear $\mathcal{H}_\infty$ control. Syst. Control Lett. **16**, 1–8 (1991)

122. Van der Shaft, A.: $\mathcal{L}_2$-Gain analysis of nonlinear systems and nonlinear state feedback control. IEEE Trans. Automat. Contr. **37**(6), 770–784 (1992)
123. Van der Shaft, A.: $\mathcal{L}_2$-Gain and Passivity Techniques in Nonlinear Control. Springer, London (2000)
124. van Keulen, B.: $\mathcal{H}_\infty$ Control for Distributed Parameter Systems: A State Space Approach. Birkhauser, Boston (1993)
125. Vidyasagar, M.: Nonlinear Systems Analysis. Prentice Hall, Englewood Cliffs (1978)
126. Wang, T.: Stability in abstract functional-differential equations. I. General theorems. J. Math. Anal. Appl. **186**(2), 534–558 (1994)
127. Wang, T.: Stability in abstract functional-differential equations. II. Applications. J. Math. Anal. Appl. **186**(3), 835–861 (1994)
128. Wang, T.: Exponential stability and inequalities of solutions of abstract functional differential equations. J. Math. Anal. Appl. **324**, 982–991 (2006)
129. Wesson, J.: Tokamaks, 3rd edn. Oxford University Press, New York (2004)
130. Westerberg, S., Mettin, U., Shiriaev, A., Freidovich, L., Orlov, Y.: Motion planning and control of a simplified helicopter model based on virtual holonomic constraints. In: Proceedings of the 14th International Conference on Advanced Robotics, Munich, pp. 1–6 (2009)
131. Willems, J.: The Analysis of Feedback Systems. MIT Press, Cambridge (1971)
132. Willems, J.: Dissipative dynamic systems, part I: general theory. Arch. Ration. Mech. Anal. **45**, 321–351 (1972)
133. Willems, J.: Dissipative dynamic systems, part II: linear systems with quadratic supply rates. Arch. Ration. Mech. Anal. **45**, 352–393 (1972)
134. Witrant, E., Joffrin, E., Bremond, S., Giruzzi, G., Mazon, D., Moreau, O.B.P.: A control-oriented model of the current profile in tokamak plasma. Plasma Phys. Control Fusion **49**(7), 1075–1105 (2007)
135. Wu, J.: Theory and Applications of Partial Functional Differential Equations. Springer, New York (1996)
136. Xiao, M., Basar, T.: Nonlinear H_∞ controller design via viscosity supersolutions of the Isaacs equation. In: Stochastic Analysis, Control, Optimization and Applications, pp. 151–170. Birkhauser, Boston (1998)
137. Xie, L., de Souza, C.: State estimation for linear periodic systems. In: Proceedings of the 24th IEEE Conference on Decision and Control, San Diego, pp. 3188–3193 (1990)
138. Xu, C., Ou, Y., Schuster, E.: Sequential linear quadratic control of bilinear parabolic PDEs based on POD model reduction. Automatica **47**(2), 418–426 (2011)
139. Zames, G.: Feedback and optimal sensitivity: model reference transformations, multiplicative seminorms and approximate inverses. IEEE Trans. Autom. Contr. **26**(2), 301–320 (1981)
140. Zheng, K., Bentsman, J., Taft, C.W.: Full operating range robust hybrid control of a coal-fired boiler/turbine unit. ASME J. Dyn. Syst. Meas. Control **130**(4), 041011-1–041011-14 (2008)
141. Zhou, K., Doyle, J.: Essentials of Robust Control. Prentice Hall, Upper Saddle River (1998)
142. Zhou, K., Doyle, J., Glover, K.: Robust and Optimal Control. Prentice Hall, Englewood Cliffs (1996)

Index

Symbols
$\mathcal{H}_\infty$ (sub)optimal control problem in time-varying setting, 49, 51
$\mathcal{H}_\infty$ (sub)optimal control problem, 12
$\mathcal{H}_\infty$-design procedure, 93
$\mathcal{L}_2$ gain, 72
$\mathcal{L}_2$ gain, 71
$\mathcal{L}_2$-gain, 55

B
backlash, 152

C
causal, 12
center manifold, 56
Coulomb model, 124

D
Dahl model, 125
Dini super-/subdifferential, 69
discrete-continuous system, 97
disturbance attenuation level, 12, 49

F
feasibility problem, 15
friction, 124

H
Hamilton–Jacobi inequality, 74
Hamilton–Jacobi–Isaacs equation, 60

Hamilton–Jacobi–Isaacs inequality, 83
Hardy space, 7

I
internal stability, 11

L
Lipschitz continuous function, 67
local detectability, 61
local exponential stability, 56
LuGre model, 125
Lyapunov function, 74

M
moving Poincaré section, 177
multilink robot manipulator, regulation of, 123

N
nonlinear $\mathcal{H}_\infty$-control problem, 58

P
periodic motion generation, 169, 183
proper/strictly proper, 6
proximal solution, 74
proximal super-/subdifferential, 68, 69
proximal super-/subgradient, 68

R
Riccati equation, 77, 79, 90, 94, 95

S

sampled-data measurement, 96
Schur complements, 15, 16, 111, 116, 119, 200, 205
semicontinuous function, 67
servomechanism with backlash, regulation of, 151
steady-state response, 57
Stribeck model, 125
strict bounded real lemma - time-varying/periodic version, 46, 48

T

transfer function, 5, 6
transverse coordinates, 178
transverse linearization, 178

V

virtual constraint, 169–171, 182, 186, 188
viscosity solution, 69
viscous friction, 124

Printed by Libri Plureos GmbH
in Hamburg, Germany

In case Publisher is established outside the EU,
the EU authorized representative is:
Springer Nature Customer Service Center GmbH
Europaplatz 3, 69115 Heidelberg, Germany

If you have any concerns about our products,
you can contact us on
ProductSafety@springernature.com

MIX
Papier aus verantwortungsvollen Quellen
Paper from responsible sources
FSC® C105338